'A momentous achievement.'

'Of all the guides to what is ahead, none is so sensitive, comprehensive and realistic as Patrick Dixon's book.'

'If you want to intelligently invent your own future, this is the guidebook. Dixon presents the best analysis of future trends.'

'Making real, significant change in organisations can take many years, so we need to be thinking deeply now about the future. That's why visioning and futuring capabilities sit at the centre of managerial competencies. This book provides an excellent and thought-provoking view of what the future may hold for organisations. Invaluable for any manager plotting organisational progress.'

Comments on the first edition of *Futurewise*:

'I was eager to see how your predictions compared with reality. Therefore I read once more [the first edition] of *Futurewise*. Result: continuous astonishment at how much of what you predicted has become true – in many cases at an even more accelerated pace than originally assumed. This result merits a huge bravo and encouragement.'

'The rising star of trends researchers.'

'I love having a model to make things develop from. The Future model is great to take hold of the paradoxes in every second of today's world.'

Comments after Futurewise presentations:

'Superb. One of the most interesting and entertaining presentations I've ever heard.' 'He deserves more time.' 'Super!' 'A real "out there" thinker.' 'He challenged many of our business beliefs.' 'Top performance.' 'Energetic, interesting.' 'Very entertaining style.'

'Thank you for your presentation at the London Business School. Naturally, you were a great hit (highest rated session on the programme). Like all the participants, I got a lot out of it and it will definitely change my work life.'

'Please express our deepest gratitude for participating in the Microsoft Global Accounts Summit last week in the Hague. I don't think I have to tell you that your session was very well received. Every person in that room was glued to what you had to say.'

'I attended the Human Resources Strategy programme at the London Business School. Your presentation on that programme was one of the most striking "events" I ever experienced.'

'I first heard Dr Patrick Dixon's outstanding presentation at a Futurists conference for around 50 CEOs and Chairmen. Other speakers included the UK former Prime Minister John Major, the editor of The Economist, Senator Bill Bradley, President Carter's former security adviser, someone from MIT Medialab, Professor Rudi Dornbusch and others. I learned more about my business from Dr Dixon than from all the others put together.'

DR PATRICK DIXON is often described by the media as Europe's leading Futurist and has been ranked as one of the world's 50 most influential business thinkers alive today.* He is a Fellow of the Centre for Management Development at London Business School, author of twelve books including *Futurewise*, and Chairman of Global Change Ltd. He advises global company boards and senior teams on the strategic implications of a wide range of global trends including the future economy, the digital society, virtual corporations, financial services, biotechnology, lifestyle changes, consumer behaviour, marketing, public policy, the future of Europe and corporate ethics.

Dr Dixon is also one of the most sought after international conference speakers. His multimedia-vision of the future is experienced by up to 3,000 people at a time, in up to four countries a week. He has presented at the World Economic Forum (Davos), the WEF Southern Africa Economic Summit and the International Emirates Forum. A past member of the World Bank Technical Assistance team in China, his clients include Hewlett-Packard, Microsoft, UBS, Credit Suisse, PricewaterhouseCoopers, Ford, IBM, Roche and GSK.

As a global authority on the future he has featured in over 150 TV and radio broadcasts in the last year, including CNN, CNBC, ABC News, Sky News, BBC, ITV and Channel 5. He has written for many publications including Time magazine and his own Web-TV station has had 6 million hits in a year, and up to 50 million words downloaded a day, running from a cyberbubble studio at the top of his house. In it he lives in the year 2010 and sees tomorrow as history.

He is also involved in humanitarian projects in the poorest nations, and is the founder of ACET International – a global network of community-based AIDS prevention and care programmes. Dr Dixon trained in medicine at Kings College Cambridge and Charing Cross Hospital in London. In 1979 he began his own IT startup in artificial intelligence and medical computing before going on to specialise in the care of those dying of cancer. He is 45-years-old and is married with four children.

* Bloomsbury Publishing/Suntop Media global executive survey: www.thinkers50.com

FUTUREWISE

six faces of global change

*A personal and corporate
guide to survival and success
in the third millennium*

PATRICK DIXON

This third edition published in
Great Britain in 2003 and
updated in 2004 by
PROFILE BOOKS LTD
58A Hatton Garden
London EC1N 8LX
www.profilebooks.co.uk

First published in Great Britain by
HarperCollins*Publishers* in 1998

3 5 7 9 8 6 4

Set in Janson and Ellington by
Rowland Phototypesetting Ltd,
Bury St Edmunds, Suffolk

Printed and bound in Great Britain by
Bookmarque Ltd, Croydon, Surrey

A CIP catalogue record for this book
is available from the British Library.

ISBN 1 86197 7107

Contents

To Sheila, my best friend,
closest adviser and source
of endless encouragement
for over 30 years.

Acknowledgements

I am indebted to the thousands of senior executives who have shaped this third edition by their participation in presentations on the Six Faces of the Future around the world and at London Business School. The content has been forged in the realities of my own experience. I am particularly grateful to Professor Prabhu Guptara, Director of Organisational Transformation and Development at United Bank of Switzerland for his unfailing encouragement and thought-provoking perspectives. I am also grateful to Simon Walker for sharpening my thinking on a number of issues including shareholder value. Thanks are also due to Jonathan Rice for help with editing, and to Shirley Bray and Martin Roder for additional research and support.

I am also indebted to a host of great thinkers, debaters, speakers and writers whose work over the years has permeated my own thinking and influenced my evolving view of the world. I owe a lot to such people as Nicolas Negroponte, Charles Handy, Lynda Gratton, John Naisbitt, Rudi Dornbusch, Peter Cochrane, Peter Drucker, Barry Minkin, David Koren and Kenichi Ohmae. David Stanley also made a very useful contribution to the method for Futuring an organisation, described in the Appendix.

I am also grateful to others who have shaped and influenced my values: Sheila my wife, my parents and many close friends including Gerald Coates and Steve Clifford. Statistics and other important data are from published government and other official sources, except where indicated.

patrickdixon@globalchange.com
Author's Web TV site – 4 million visitors:
http://www.globalchange.com – latest updates on issues
January 2004

Introduction

Life in the third millennium

You cannot fight against the future. Time is on our side.
W. E. Gladstone 1809–98

Either we take hold of the future or the future will take hold of us.

The third millennium began with a big party, which rapidly became a huge hangover with the crash in dot-com stocks, the American economic slowdown, corporate scandals, attacks of September 11, Afghan and Iraq wars – and all that followed. It was a vivid example of why we urgently need the big picture, to plot the longer term sweep of future history, or lose focus and direction.

We need to be futurewise. That means planning to change tomorrow, future-thinking at every level, taking a broad view to out-plot the opposition. Being futurewise is about more than mere predictions, it's about shaping the future, making history, having contingencies, staying one step ahead. This is an extraordinary time to be alive, at the start of a new millennium. The world is being transformed before our eyes from an emerging industrial revolution and a technological post-war society into something altogether new and different. This millennium will witness the greatest challenges to human survival in human history, and many of them will face us in the early years of its first century. It will also provide us with science and technology beyond our greatest imaginings, and the greatest shift in values for over fifty years.

This third edition of *Futurewise* has been altered very little from the vision of the future first set out in 1998. Many of the things that I expected are already history, such as: the threat of tribalism; the growing protests against globalisation; the progressive impact of digital technology; economic instability; unsustainable pace of mega-mergers; the growth of 'reality' TV shows; chips fused with brain tissue, spectacular advances in human cloning, global agreement on carbon trading; rioting in France over pensions and strains in Europe following the Euro, prior to enlargement.

None of the many expectations listed in earlier editions have been removed, unless they have now become history, although some timings have been adjusted forwards. We may debate about dates but the trends on which *Futurewise* is based are clearer than ever. So what is coming next? While no one can predict the future, there are fundamental processes at work which have many consequences. From these we can plot out reasonable expectations – things that could happen which need to be considered and prepared for. That is the futurewise challenge.

SIX FACES OF THE FUTURE

The future has six faces, each of which will have a dramatic affect on all of us in the third millennium. Each is important but not equally so, depending on who you are and where you are positioned on this globe and on the social scale. It is impossible to keep them all in view at once: some are related, others are opposites. Together they form the faces of a cube which is constantly turning. The faces spell the word 'FUTURE'.

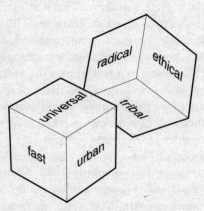

Fast and Urban sit together on one side, Radical and Ethical on the other. On top is Universal, and beneath at the opposite extreme is Tribal. Most executives spend their lives looking at the cube from above, at a world which is fast, urban and universal. However, one twist through 180 degrees presents us with a very different view: a world which is tribal, radical and ethical. Understanding the tension between these two dominant views is essential to understanding life in the third millennium. As we have seen, a tiny minority who are strongly radical, ethical and tribal can affect the rest of us profoundly.

Expectations, predictions and challenges to management

Out of these six faces cascade over 500 key expectations, specific predictions as logical workings-out of these important global trends. These range from inevitability to high probability to lower probability – but still significant enough to require strategic planning and personal preparation. There are more than 200 challenges to management – key questions which demand answers. There are more than 100 futurewise issues for individuals. However, there is one single overriding factor which is central to understanding tomorrow: the Millennium Factor.

THE M FACTOR

Few people have woken up so far to the impact of the millennium. My children are the M generation. Their entire adult existence will be lived in the third millennium. Children born today have no memory of the second millennium. The M Factor will not be instant but it will be profound, far-reaching and very long lasting. Expect to see the M Factor affect every aspect of life on earth over the next 150 years.

We are seeing it already in many countries, as a radical rethink about values. Indeed, whenever I talk to people about the future, they talk to me of their concerns for themselves, their families, their communities and the whole world.

Making sense of our history

The human brain makes sense of the past by dividing it into intervals: the day marked by the sun, month originally by the moon, year by season. Then there are decades and centuries. So the nineteenth century becomes the Victorian era, and is seen as a single defining period with its own distinct culture and traditions. But unlike the sun or moon cycles, these time-stones are entirely artificial, set in concrete only by the diary of humankind. They are entirely the product of a human need to pigeonhole events into neat time-frames. And four time-events were to hit us in the same instant. New year, decade, century and millennium.

Every decade has a character

Every decade has its character. Only a millisecond of eternity separated the sixties from the seventies yet we all recognise instantly the music, style and architecture of the sixties. The same could be said for every decade in the last 100 years. It is totally irrational to think that whole periods of human existence can be neatly framed by decades, centuries or even millennia dated from the hypothetical birth date of Jesus Christ, but they are. We all know what we mean when we say that a building is nineteenth-century. By the year 1904 people recognised that the old century was dead.

Get into the third millennium

In the latter half of the twentieth century if you wanted to insult your boss you would have said that he or she was still stuck in the seventies or eighties – or perhaps even the nineteenth century. However, the insult has changed: 'You're still stuck in a late twentieth-century time warp. Get real, this is the third millennium!' or 'That's so last century'.

No architect wants to design a late twentieth-century building. At Greenwich in London, a huge millennium dome, designed by Richard Rogers, was built to celebrate 2000. Imagine a group of Japanese tourists being shown round in 2050. What will the guide say?

There is only one accolade that is likely to satisfy architects of the large number of great buildings opened in the early 2000s and that is: 'Ladies and gentlemen, here is an outstanding example of early third millennium architecture, expressing as it does the hopes, aspirations, dreams, anxieties and fears of the new millennium.'

Expect to see hundreds of millennium structures around the world, all curiosities by 2010, many decaying and an embarrassment by 2025, all competing to be the defining image of a truly third millennial building. Expect more Eiffel Towers: big white elephants criticised at the time that people grow to love and then fight to save.

Expect third millennial fashion, clothes, radio, television, culture, music, art and social codes. The real winners will be those who tap into this huge shift – and help define it. What television producer will want to produce second millennial TV? What clothes designer dare risk his annual collection being labelled as a rehash of tired late twentieth-century fashions? Every creative talent will be focused on trying to interpret what the third millennium means. Expect to see radical shifts by 2020 in every aspect of art and culture with eccentricity pushed to the limits of every extreme, before settling down into a third millennial rhythm of life.

Changes can be dramatic – look at the shifts in social customs and dress from the eighteenth to nineteenth to twentieth centuries. Do we really believe the globalised dark suit and tie will still be standard uniform in 2050? Expect not only major shifts in fashion but also revolutionary new fabrics.

So what does third millennial life look like? Faster, more technology-dominated, data-obsessed but more intuitive, sensitive and environmentally aware.

If you are a pre-millennialist . . .

Pre-millennialists tend to see 2000 to 2010 as just another decade. The trends of the eighties and nineties continue, just more of the same. Post-millennialists are very different. They are products of the third millennium. They live in it. They are twenty-first century people, a new age. Expect to see one of the greatest generation gaps in recent history between pre- and post-millennialists. The trouble with trends is that we always look forward. We look back to find the line of the curve but our perspective is blocked by the patterns of the past. Looking ahead is progress.

However, the lesson of history is that the pendulum always swings. It is never still except for a millisecond at the outer limit of each swing. It is true that at extremes the pendulum moves relatively slowly and it is far harder to tell the current direction. It never swings true, but always twists somewhere new.

Trends and countertrends

To every trend there is a countertrend, which is why media pundits are able at once to describe, for example, trends to greater liberalism and greater conservatism. Both are true – of different tribes in the same society. So gay rights movements continue to make advances at the same time as America is becoming gripped by cultural conservatism, with ideas such as marriage, religion and civil society seen as the answer for the future.

Drug use soars, with growing calls for decriminalisation, at the same time as a neo-prohibitionist movement seeks to make it all but impossible to smoke a cigarette in a public place. Expect to see millennial culture clashes between opposing trends, a world increasingly of extremes with tendencies to intolerance as groups fight to dominate the future. But not driven merely by *culture* clashes, the greatest forces will be unleashed by clashes of *conscience* influenced by religious conviction, or lack of it.

Which trends will be dominant?

The big question is this: if trend and countertrend coexist, which will be dominant in the new millennium? The truth is that in a pluralistic, multi-track society there are a number of pendulums operating. Dominance is less important with the emergence of micro-communities, micro-markets where all that matters is being able to target every trend with a package of products and services. Expect to see whole industries built around micro-marketing techniques, micro-advertising, micro-distribution networks, micro-affinity groups.

Trend 'wild cards'

Expect trend 'wild cards' over the next twenty years and plan for them with rapid-reaction capability and streamlined decision-making: huge unpredictable events such as bio-terrorist or dirty bomb attacks, war, a nuclear accident or the unplanned launch of nuclear weapons, vast volcanic eruptions or plagues or even a comet collision with enormous destructive power.

Inside the mind of a post-millennialist

The key to understanding the post-millennialist is one word: sustainability. The reason is simple: current trends are unsustainable, or seem to be. We have never before had the means to view an entire century and the effect is extraordinary. One hundred years of film – still and moving – has charted almost every detail of our lives. We began with horses and carts and ended with people living in space. We began with books of paper and card and ended up with cyber-reality. Can we survive another hundred years of increasingly rapid change? What about another thousand years? Could this be the last millennium?

Economic growth is at the centre of every government strategy, yet in the third millennium expect that economic growth as a goal will be increasingly questioned. Growth means more things, greater wealth, but does it mean greater happiness? If quality of life means being happy and fulfilled, what is the secret of it all? Since multi-billionaires show little or no sign of being any happier than lesser mortals this is a fundamental question.

Expect attitudes to change. They are bound to, since the premise on which this whole edifice is based is so flimsy and irrational.

Michael Svennevig, Research Director at the Centre for Future Communications at Leeds University, has identified four new social groups.

- Embracers – people not unlike eighties 'yuppies', living in warehouse flats with the latest gadgets, surround TV and computers which they see as signs of success. They have low social conscience.
- Pragmatists – care about the community and watch less TV to relax. They favour technology for its human benefits.

- Resistors – more than a third of the population who feel life is passing them by. Insular, homophobic, uncharitable, heavy-smoking.
- Traditionalists – a seventh of the population. Mixed feelings about technology. Happy and self-confident.

Each is a market sector. Each has its own set of pendulums. Each will react to the events described in this book in different ways.

You may be an optimist or a pessimist. The future is uncertain and many possibilities are alarming, but I am an optimist: convinced by the potential of science, medicine and technology, and the capacity of human beings to build a better kind of world. But the way ahead will not be easy, beset with moral challenges, economic and geopolitical instability, together with resource limitations.

Radar Screen and Road Map

So then, here is a radar screen and a road map. The radar screen is a grid you can place over any organisation, a structure for thinking about the future. It will help you scan your own world for emerging trends.

Most business time is spent managing high probability, low impact events, which are usually extensions of existing business activity. But the really interesting areas to watch are the small blips on the outer edge of the radar screen, some of which are moving rapidly. These low probability high impact events can strike rapidly and transform your world. It's easy to dismiss them as improbable and therefore insignificant. However there are a great number of them. So the risk of your own business being hit by such a low probability, high impact wildcard is far higher than you may realise.

The road map? You will find in this book a way not only to make sense of new blips on the radar screen but a guide to how they may behave in the general frame of the future.

CHAPTER I

Fast

Speed will be everything

The future is something which everyone reaches at the rate of 60 minutes an hour, whatever they do, whoever they are.

C. S. Lewis

In today's fast-paced culture, ridden with sensory overload, kids are exposed to an input beyond the comprehension of those who lived a century and a half ago.

US Juvenile Court Attorney, August 1997

HISTORY IS ACCELERATING

The first face of the future is FAST: speed will be everything. Never before has the future so rapidly become the past. History is accelerating whether you look at trends in the economy, global events, industry, social factors or politics, or dotcom hype and bust.

If the nineteenth century was the age of the large-scale machine (steam engines, power looms, vast factories), then the twentieth century was the age of the petrol engine and information technology. We talk about rapid change today but we forget the speed of change in earlier generations. The outbreak of war in every century has tended to create sudden, overwhelming social and economic changes, as trade routes became blocked or people moved to avoid conflict, or found their homes and property suddenly destroyed.

Look what happened with cars. Henry Ford formed his automobile company in 1903 and sold his first car that year. It was an amusement for the wealthy: slow, unreliable and expensive. Within ten years the first assembly line was running, a complete revolution in manufacturing. By January 1927 there were 9 million cars in the US, up by 8 million since 1924. That was the take-off point. Until 1924 cars were expensive toys. By the end of 1936 they were near necessities.

Air travel also developed faster than many realise.

Development of Air Travel

- 1903 First flight – a total airborne distance less than the wingspan of a Boeing 747
- 1919 First transatlantic non-stop flight
- 1931 TWA first air freight service
- 1954 Boeing four-engine 707 launched
- 1959 Flights at twice the speed of sound

What started as a dangerous hobby for the half-mad, with 100 deaths by the end of 1911 – a high rate among very few fliers – soon became the deciding factor in a world war, which hung on who dominated the skies.

Political whirlwinds affect whole continents

Look at the speed with which the Soviet Union collapsed in 1990. When the Iron Curtain fell people thought the reunification of Germany would take five years. It took five months – although with huge longer-term problems. Expect further rapid realignments, with North Korea top of the list as the last outpost of Stalinism, a country bankrupted and starving after three decades of mismanaged central planning and its intense suspicion of all neighbouring countries. North Korea could crash at any moment, spilling thousands of starving refugees into China, South Korea and Japan. Expect, too, increasing signs of regional instability in China.

Look too at the rapid creation of the global alliance against terrorism, formed in less than a month in 2001, with unheard of co-operation and international consensus, Russia and the US stood shoulder to shoulder in warm solidarity – until the Iraq war.

Trends are becoming more unpredictable

Take Mexico's financial crisis in 1995, dubbed the first financial crisis of the twenty-first century. It hit with ferocious speed, as global investors fled from the tumbling peso. Then came Thailand's devaluation of more than 23 per cent in 1997, following huge exchange rate fluctuations which some speculated were added to by drug-related money laundering flows. Thai authorities wasted £19 billion propping up failing financial institutions exposed to bad loans, before shutting down 42 companies. The Bank of Thailand lent 10 per cent of the country's entire gross national product to 91 finance companies. Then came the sudden collapse in the currencies of the Philippines, South Korea and Malaysia, along with 70 per cent devaluation in Indonesia, and Turkey's 50 per cent devaluation in 2001. Expect further runs on currency in emerging economies and a rush into protective alliances which will also be overpowered by market forces, driving both inflation and deflation in different sectors.

Tigers turn to lion cubs

The so-called 'tiger economies' of South East Asia grew fast on the back of cheap labour and cheap exports but now have the whole of China to compete with, while the expectations of their own labour forces have risen.

New contingencies, with IMF and inter-bank co-operation, better investor-information and better government communication, are not going to be enough to prevent more speculative attacks on one currency after another. Global money flows are just too big to control. Even the Bank of England was beaten by the market when it ditched the Exchange Rate Mechanism, heralded at the time as the right next step towards European integration.

The collapse of the ERM was linked in the minds of the ignorant with speculators like George Soros who made a fortune that week. He and others have similarly been blamed by senior politicians such as Dr Mohathir Mohammed, Malaysia's prime minister, who called them 'rogue speculators' wanting to destroy weak currencies. He said speculators 'should be shot' and lamented that 'most of the work we have done' in developing Malaysia over the past 30 to 40 years had been undone 'in a period of two to three weeks'. His comments followed a devaluation of less than 10 per cent.

Expect increasing North–South tension as emerging economies come to realise that abolishing all trade and currency restrictions in a rush for growth also places their countries at the mercy of rumours, hunches and market opinion. Expect an even greater backlash against globalisation, with some nations reduced to 'economic slavery' by massive, destabilising currency flows. Expect large institutions to continue to make (and lose) huge fortunes trying to outguess volatile markets in these countries. Expect countries to rally round to help stabilise each other's currencies, as seen in Thailand where China, Japan, Australia, Singapore and Malaysia were among those contributing emergency loans. Expect far more countries to see rioting in the streets as workers, students, wealthy intellectuals and the retired all unite to vent their various angers and frustrations at leaders, global institutions, wealthy, 'arrogant' nations and ethnic minorities.

World markets will be seen more in the future for what they are already: a great global gambling den, using identical skills to betting on horses: hunch, intuition, detailed analysis, inside knowledge, a host of other factors – and a dose of good fortune.

Instability of basic commodities

Basic commodity prices will also continue to fluctuate wildly at times. Take zinc for example, whose price fell 18 per cent in an hour on 29 July 1997, catching on the hop Chinese producers who had pre-sold what they did not own. Expect increasingly complex investment instruments to be developed, so that a commodity some-times rises or falls dramatically as a large market intervention is made, linked to a completely different and apparently unrelated event. Expect growing worries about these derivatives.

Management gurus are the high priests of confusion

Every week there are more books on management. Each one often contradicts what has gone before, struggling to find a fresh view. Insecure managers continue to gobble up the latest fads, kissing common sense and their own experience goodbye. Expert, confident managers will continue as before to dismiss management fads in favour of their own intuition and intelligence, working out their own solutions, adapting and borrowing as they go from a wide variety of sources.

However, the speed of change will guarantee an almost permanent supply of semi-neurotic managers who are constantly on the lookout for some new, comprehensive solution to their day-to-day problems. But where is the real evidence? Anecdotes, one-off case studies and personal opinions are not enough to run a business by. They are no substitute for rigorous analysis.

Management theory is still an immature, inexact and unproven semi-science. Expect that to change over the next two decades as rigorous statistical and analytical tools are devised to prove or disprove the key elements of success in management methods.

Expect 'management historians' to become sought after, analysing industrial successes and failures during the previous Industrial Revolution and at the turn of the twentieth century.

Better 'early warning systems'

The interval between early signs and a full-blown new trend is shorter than ever and long-range forecasting is becoming more difficult. The narrower the field, the harder it is. So, for example, while the trend towards global networking is undisputed, the exact year in which online share trades will really start hurting the profitability of traditional brokers is not.

This means that corporations need to have far more sophisticated early warning systems, able to tell the difference in a graph between 'background noise' (minor changes) and the first sight of a major new trend. The trouble is that most decision makers tend to be cocooned by people in the same industry or even the same company. Corporate blindness and industry blindness are real dangers. The most important new trends may be most obvious to experts outside the work and culture of the institution. An example is Internet banking, where banks, very wisely, have drawn heavily on consultancies from high-tech companies rather than just from financial services experts.

Being futurewise means thinking laterally, taking the broad view, considering the 'wild card'.

The problem with large institutions is the time lag between a board decision and the mobilising of the entire company in a new direction. Just drawing up plans for approval can take months. But today's world requires a different approach.

Multi-track scenario planning is needed, with plans laid out for several options and some investment in each. The extra cost is offset by the extra profitability gained through being able to move faster than the rest.

The future will not be more of the same

Today's technology is tomorrow's dinosaur. The trouble is that most companies are obsessed with pushing accepted technology to the limits, when the greatest long-term threat is probably going to come from a technology which is new and different – so different that very few can take it seriously today. At the same time, most individuals struggle to keep up with technology they already have, and are unable to grasp the full impact of the future. They stumble from one new software application or tool to the next: a new e-mail system, a word processor upgrade, a new telephone switchboard, a new manufacturing process. Success means the fast integration of today's new technologies and preparation now for the next generation of tools tomorrow.

The lesson of history is that companies usually fail to cross the bridge from the old to the new: they just shrivel and die, driven by people who are future blind. Harvard Business School has described 'discontinuities' that occur when new technology wipes out all the existing players in an industry.

An example from a hundred years ago is ships. Sailing ships ruled international trade. When steam power came it was regarded as a joke for powering large ocean-going ships. Steam boats were too small and too expensive. Established shipbuilders ignored the technology, while new companies seized the initiative. When steam power improved, the old companies tried to catch up but it was too late. Not one shipbuilding enterprise in the US managed to make the transition.

Discontinuities leave companies behind

The same happened with hard disk manufacture for computers. The original large disks were fast and cheap. Along came a new format, smaller, slower and less reliable. Customers knew they wanted what they already had, and could not see the point when the new disks were first made. Established disk drive manufacturers laughed at the new technology and ignored it. Staff left to set up new companies which later took a big slice of the market. By the time the old producers realised what was happening they were too late. The performance of the old drives had continued to improve – but faster than customers needed or wanted. Meanwhile the newer technology had caught up – and was cheaper.

Don't believe market research

Business tends to see through the lens of what it thinks it can sell but customers only know what they are used to. Customers don't understand enough, nor are they visionary enough to be able to guess what they might want from a totally unknown technology in the future. Market research only tells you how people think and feel today. It tells you little about the future beyond tomorrow.

If a bank had asked its customers in 1995 whether they would want to spend hours at home staring at a computer screen, managing their own accounts, and buying books, food or CD-ROMs, most would have said no. In fact most of them would have been shocked at the question. Along came the new, improved, graphic-rich Internet, providing instant global access. Within a couple of years millions of people were finding unthought-of ways they wanted to use the technology, including buying stocks and shares and managing their accounts. Some experts saw it coming but the average customer completing a survey did not.

Another market research failure was WAP: the cut-down version of the web for mobile phones. People told researchers it all sounded great – but take-up was very low because screens were too small, keyboards were fiddly, data was slow and decent WAP pages few. Market research also failed to predict the dramatic growth of short text messages (SMG) which reached 1 billion a month by the end of 2001.

Expect consumer surveys and market research to be sidelined by futurology-based customer profiles. Market research only tells you what people want today. What's so smart about that when planning a new range of products for tomorrow? Market research is non-exclusive data. Anyone can go out and ask the same questions, so where's the real competitive edge? Everyone has to do it for feedback on what's happening today, but the faster the world changes, the less relevant market research becomes for planning (see also pages 178–179).

THE TELEPHONE REVOLUTION

In 1977 the latest technology allowed 24 telephone conversations to be transmitted on four copper wires using multiplexing. But by 1997 around 70 million conversations could be carried on just two optic fibres, each not much thicker than a single human hair. Telephony had become almost free. Capital costs are high but once those fibres are laid, light is cheap and fast. And that was before the Internet. And the next decade?

Expect to see widespread availability of global calls for a flat rate, regardless of distance. One reason for this is that measuring the time and distance of every call is so expensive as a proportion of total call costs, which are falling dramatically. Expect most households in wealthy nations to have several phone numbers by 2006. Expect actual numbers of phone lines to be meaningless by 2010, as multiplexing allows anything from two to a hundred different data streams to be combined down a pair of copper wires, or tens of thousands down a single co-axial cable. This means that most executives will have access to far more telephone lines at home than they do at work today for their personal use. And flat-rate calls to the other side of the world will give a huge boost to home-based virtual working. At the same time, high quality multi-channel interactive TV, video links and data streams will alter social behaviour and family relationships. It's already happening. When travelling I often connect my hotel room to my home for days at a time – with sound and video – using the broadband net connection. I sit virtually in my office at home. Our children are able to walk virtually into my hotel room. Hotel charges range from zero to ten dollars a day.

Mobile telephones replace landlines

In 1984 experts predicted that by 2000 there would be over 900,000 mobile phone users in the US alone. But by 1997 there were already 40 million. Mobile telephony is also becoming free. In Britain people have started using mobiles in all kinds of bizarre ways, now that they can use them as long as they like at certain times for free. One company offered unlimited local mobile calls (to 6 million people in London). People have started using them for babysitting.

A couple are about to go out for the evening to a wine bar at the end of the road. They take the mobile and phone themselves while still in the house, then fix the other phone to the baby's cot. With the mobile held to an ear, they wave goodbye and close the front door. Walking down the street, they can hear every sound. They enter the wine bar. Every few minutes they listen. The call has cost them nothing. Irresponsible of course, but this is what happens when wireless calls are free.

Expect enthusiastic post-millennialists to abandon landlines for mobiles in most circumstances by 2005. Expect many office switchboards to use any landline or mobile as extensions, creating virtual global telephone exchanges. Expect all except disposable mobiles to have cameras and colour video screens. Expect growing worries about phone radiation causing health problems – and many lawsuits, although risks to individual users seem very low.

Emerging economies leapfrog over old technology

Developing countries have leapfrogged over old copper networks with the latest mobile telephony. Just look at the numbers of people with mobiles in Delhi, Calcutta, Beijing or Moscow. Mobile technology allows a developing city to install instant city-wide 3G networks with just a few masts on buildings. Microwave and other technologies have replaced trunk cabling in places like South Africa – they are more reliable in places where copper cables are stolen to order by big gangs.

Expect a continued boom in wireless technology in all emerging economies. Expect the collapse of government monopolies in telecommunications in many African nations, their power broken by millions of satellite phone users. Expect national governments to become less anxious about the foreign control of telecoms.

The ideal phone

The ideal phone weighs almost nothing, has batteries that never need charging (miniature fuel cells running on methanol), accepts voice commands, and works anywhere – even in a 20-mile tunnel. It's coming. The next decade will see the universal use of mobile devices in our pockets which switch when needed to landline or to satellite or to local wireless networks.

Expect phones to get larger and smaller. Larger for those wanting full integration with personal organisers, videostreaming and web surfing. Smaller for those in a voice-activated wireless world. Expect manufacturers to stumble with many losing their way as old-style hand-held mobile phones are swept away by micro and macro devices, as well as PCMIA mobile cards for laptops and a host of other 'cross-over' applications, many with inbuilt global positioning.

Call centres are the new sweatshops or workhouses

Expect a new push in telesales, call-centre boom replacing face-to-face retail sales. Staff will be located anywhere in the world. You are only as good as you sound. Accent is the key. Scotland and Wales are boom areas for UK centres because surveys show people trust those dialects. Call British Airways in London and you will be answered in Newcastle. Call American Airlines in Paris and you will be answered in New York. Call the local town hall in south east England and you will be answered in Scotland.

It is high-pressure, chicken-coop work with 90 per cent of employees' time spent on calls, crammed into little boxes in front of a computer screen. Expect telebusiness to mature with a well-recognised career path of its own. Expect new regulations in Europe and North America concerning the working environment of these humans working in battery-hen conditions. Expect fierce competition from well-established call-centre labour forces in emerging countries such as India, where labour costs per call are 80 per cent less and phone charges almost identical.

Expect a big reaction against companies that go on forcing callers to push buttons in response to endless choices and a return to human beings on the end of the line. People need more contact not less in a virtual world. Expect voice recognition to allow key words to be understood from near-continuous speech in replies by 2005, regardless of accent.

Expect the near-universal implementation of intelligent call-answering, with all incoming lines switched automatically to the person or department which last dealt with calls from that number. All operators will be expected to have complete data on the customer on-screen seconds before they answer the call, as well as computer-generated suggestions on products to cross-sell.

Cross-selling makes the margin

Expect high-volume cross-selling operations in every large business, with loss leaders used to sell higher-margin products. Expect cross-selling to diversify. For example, an insurance quote will be followed with 50 per cent discounts on weekend breaks and 10 per cent off car dealer prices.

In future you will find companies have an uncanny ability to tell exactly what you want just before you realise you need it. It's already started on the Internet, with adverts that pop up on the screen prompted by your interests and recent page requests. No more 'wasteful' mass advertising – but accurate information positioning.

THE COMPUTER REVOLUTION

More microprocessors than people

We live in a world where there are now more microprocessors than people. The power of PCs has been doubling every 18 months for many years. At that rate it means we will see machines in 2024 with processors that are 10,000 times more powerful than the fastest chips we have today. Expect breakthroughs with nanotechnology within 20 years, allowing chips to shrink massively to microscopic levels. Expect distributed processing to allow corporations to build vast number crunchers using spare capacity in thousands of desktop PCs.

Death of cathode ray tubes will be delayed

Future generations will be utterly amazed when they look back at today's screen displays. Despite vast steps in technology, in the early third millennium almost all computers worldwide were still using huge, heavy displays made with overblown valves, with ultra-high voltages, huge power consumption, poor visibility, low resolution and limited size. Slow progress in screen technology will continue to hold back the digital revolution beyond 2005.

Tomorrow's devices will be of four types, mainly flat screens.

- Miniature displays that can be magnified, for example phones, watches, glasses
- Flat displays on walls, desks and any other surface up to three metres by two metres, also for use in cars, aeroplanes. Expect paper-thin display sheets by 2005
- Projected displays onto larger areas
- Direct low-energy laser projection onto the retina.

Expect new devices which use tricks to persuade each of our eyes that they are receiving a different image. This will be the key to large-scale three-dimensional imagery. Expect digital glasses, not only correcting short or long sight but also providing a massive virtual see-through three-dimensional computer screen in front of the wearer. It will be similar to head-up displays on fighter aircraft.

Death of old photographic and film industries

The twentieth century was the age of photography. It began with a huge burst of creative still pictures, and then with movies. It ended with the imminent death of light-sensitive chemicals, papers, developers and printing.

Expect huge improvements in the resolution of data screens, projectors and cameras, rivalling finest-grain photography by 2010. Expect data cameras to replace 35mm film cameras for most domestic use by 2010. Expect ultra-high definition TV cameras to replace film in most situations by 2005, with the exception of cinema films and ultra-high quality images for advertising. But even in the film industry, digital cameras will be used increasingly, image-enhanced for higher resolution where needed for cinema viewing, employing the same technology used today in special effects generation. Since computer-generated images will be central to tomorrow's film making, it makes sense to start with digital images in the first place.

Expect tens of thousands of redundancies in the photographic and film industries. The days of silver halide on paper or celluloid are numbered.

Software will always be full of bugs

Desktop computers today are so powerful that even if technology stands still it will take the world's programmers at least 20 years to exploit their capability to the full. The trouble is that they have less than 20 months – because by then a new generation of machines will be around. More memory, different structure and a different operating system. Sure, everything is always designed to be upwards-compatible, but compatibility is a myth. Take Pentium PCs, designed to parallel process several instructions at once. The moment they arrived, the programs written for old chips were dead: slow and not worth improving.

So a brand new code was written for Pentium chips. The bugs were never sorted out in the old versions and bugs in the new ones never would be either, for the same reason. That is why the basic Windows operating system made by Microsoft was full of bugs and every upgrade continued to be bug-ridden.

The operating system was untidy and not properly documented. The largest available manuals were inadequate and confusing – despite huge efforts. The future will be even more complicated. This is worrying when you realise how much of the world's industry depends on networks of these machines for daily survival.

The situation is made far worse by the success of Microsoft and the general culture of the industry. Car manufacturers would never get away with launching cars they knew on launch day were completely unreliable, breaking down every few hundred miles. But that is normal practice throughout the computer industry, and will continue to be so for the next ten years until software becomes fully self-repairing. Even then, expect further problems.

Virus attacks will continue to soar, spread rapidly via the Internet. By 2012 all large businesses in the US were being hit on a daily basis. Expect many more dangerous viruses to emerge and expenditure on anti-virus software to increase rapidly, with weekly updates for all vulnerable high-value systems. Expect (unproven) suspicions by 2005 that some of these viruses are being created deliberately by companies selling anti-virus products, to keep a multi-million dollar market alive.

And then there is junk e-mail. A few dollars will buy 100 million addresses, generating billions of adverts a month, all sent at zero cost. 30 per cent of this junk is pornographic – some containing obscene images illegal in many countries. Yet children with e-mail addresses are also targeted. Junk wastes time, violates personal values and corrupts the innocent. Expect legal controls soon.

A world over-dependent on PCs – with problems

Almost all computers today are sold with drives containing more than 15 billion bytes of data, yet with backup devices which pack less than 2 *million* bytes per floppy disk, or less than 600 million bytes per writable CD. This is suicidal, verging on the criminal. Every user is at serious risk of catastrophe. A full backup of everything on the hard drive is essential, preferably daily, so that in the event of disk failure, software problems or theft the entire system can be recovered. But how can you possibly back up onto 15,000 floppy disks, or onto 25 CDs? How would you survive the theft of your machine? What if the thieves also stole your backup disks or tapes? Our globalised society has plunged headlong into the computer age with very little forethought.

Software itself can destroy a machine in seconds.

Disaster recovery will be a major headache

Computer sales have increased exponentially, so most computers are relatively young. Therefore the majority of big failures are around the corner. And most of them will hit small to medium sized companies the hardest. Since they comprise most of the economy, this is a big time-bomb waiting to go off. At the same time, business dependence on computers is also increasing exponentially with, for instance, the complete abandonment of paper records for day-to-day accounts systems. Yet many small to medium sized companies have no in-house expertise, and no support apart from the local computer supplier.

Half of all companies fold following a major loss of data. Expect PC disaster recovery to become a key issue, with lawsuits taken out against companies for selling backup systems which don't do what they say, that is restore a complete working system at the touch of a button.

If companies are vulnerable, then many teleworked executives at home are especially at risk. How many back up all major changes every day? How many would survive a burglary or fire? One professor recently left his laptop on a train – losing several years of research data, because his only backup was on a floppy disk contained inside the same machine. Every home worker using computers needs to take backup, bugs, crashes and viruses seriously. Expect a whole new support industry for home workers including telephone support, remote PC configuration and software repair using networking, and (rarely) same day site visits. Expect home workers to back up data online to computers in other cities as the ultimate security.

THE NETWORK SOCIETY

The digital economy is all about competing for the future, the capacity to create new products or services, and the ability to transform businesses into new entities that yesterday couldn't be imagined, and that the day after tomorrow may be obsolete.

Don Tapscott, *The Digital Economy*

In 2003 networking was still in its infancy despite all the net hype of 1999 and the bust of 2000/1. These are the earliest days, equivalent to the last years of the nineteenth century when people experimented with electric light, electric motors and the internal combustion engine. Each dominated life in the twentieth century after mockery by experts on the future. The Internet is exactly the same.

The Internet, first designed as a Pentagon creation to enable the US military to survive a nuclear holocaust, had a membership by early 2003 of some 400 million people with more than 4 billion publicly accessible documents. We can debate the speed of uptake but it is clear that by 2010 more than a billion people will be networked or have some access to whatever has replaced the old Internet. Internet power means total access at home, in the hotel, in cars, on trains, in the park, on the beach or during flight – perhaps even in sleep.

The Net combines telephony and computer power, a devastating double-act which will, step by step, transform every aspect of life in cities and rural areas. No city, town or village will remain untouched. Every home in many developed countries will be net-worked with global intelligence via interactive digital TV. Lowest income groups will also find the Net wandering in through the front door, in every phone handset, video recorder and other devices.

When first created the Net was primitive. It is still primitive. It will take at least another 20 years of mega-fast technology before it even begins to deliver some of its greatest impacts. Forget net-thinking. The net itself is a last-century idea. Life beyond the net is a world where everything, everywhere is totally, wirelessly, connected, all the time. The shift from e-mail to chat, voice calls to text and video is a foretaste of an always linked existence.

Net threat to Inland Revenue and national sovereignty

But how do you tax this cyber world? Where do you tax it? In what country's jurisdiction is a software house selling games down-line from websites housed in five different countries? If tax laws are applied in the country housing the server computer, then within weeks expect to see massive data flight, a data drain from countries imposing taxes to ones which remain free of Net taxes. If the tax is decided by the location of the software company itself, then expect immediate relocations.

Recently I decided to add to my website a bookshop offering for sale 1.5 million titles. It took less than 30 minutes to create this virtual branch of an existing online store, earning me 8 per cent commission on every sale. My company is in the UK, the web server is in the US, the online store is owned by a US company. So where is the business to be taxed?

As international regulation tightens, expect to see some countries emerge as cyber-havens with a deliberate policy of non-co-operation with international Net agreements. Expect other countries to create free zones or semi-states, geographic areas where companies can relocate as Net enterprise high-tech villages, immune from normal taxes. Expect many nation states to retaliate by seeking to punish information service providers who handle content from these non-regulated sites.

Companies will relocate not only to avoid tax but also decency laws. In 2003 pornography sales accounted for perhaps 10 to 15 per cent of all Internet retail turnover. Expect the sex industry to be a significant driving force in Internet commerce throughout the next 30 years, pushing out the boundaries of new technologies such as video phones, interactive TV and virtual reality.

The Internet has the power to break national economies, to control government policies and to redesign national frontiers. The Net will mean the end of income tax as we know it. Already we can switch unlimited amounts of money through the Internet without any trace, and send virtual cash in attachments. Smart card technologies mean yet further problems for investigators.

Take a scenario: an employee comes to me to be paid. She offers me her smart cashcard which contains a chip. Two hundred million people had these by 2003, controlling bank balances, holding cash, and a host of other information. I swipe the card through a slot in my PC and instantly transfer from my account a sum of money which she can spend anywhere in the world. All the transactions are in code. The transfer from my bank to my PC and from my PC to her card are all in secret form, so that the best detectives in the world will never know for months or years what I paid her – if anything. Bad news for the banks.

The other day I shocked the CEO of one of the world's largest banks by handing him a floppy disk with a slot inside it able to hold a smart card. Every PC in the world can become an ATM machine, replacing billions of dollars of investment in hole-in-the-wall banking machines. Value can be stored on PCs or on cards or on mobile phones and held or transferred without using banks at all.

Thus the only people paying income tax in the future could be the poor and near-destitute, because everyone else will have the power to hide income. But what happens to a sovereign state when incomes can no longer be tracked or taxed reliably? Expect a shift by 2020 from income tax to property and purchase taxes, which are harder to avoid in a virtual age – except for goods transmitted online, such as software.

Every high-value electrical device will be wired

Nandos in South Africa runs a world chain of chicken-burger outlets. In their own country they are a threat to McDonalds. They have hundreds of restaurants, globally, and big vision. Networking keeps the show on the road. The chairman can watch burgers being cooked in London from his desk in Johannesburg. He can tell the temperature of the oven (too high/too low), the number of meals cooked, the money taken, the time from order to meal and most important, the last time the ovens were cleaned. Every one of his outlets is modem connected. Robots in Johannesburg can process and filter the information, so he only sees the trends and results of their intervention.

Homes of the future are intelligent and resource saving

Expect the washing machine to call an engineer when it starts overheating and the garage doors to open as you approach, the lights to go on and the coffee machine to begin pouring a fresh cup of coffee, waiting for you in the kitchen.

We take for granted today the creeping robotisation of the world we live in. When I walk into the Gents at Zurich airport the urinal knows I'm there and when I've finished. So does the hand-drier, and the exit door. Link these together with other sensors and devices and we have the beginnings of an intelligent environment.

Expect all new homes in Western countries to be intelligent from 2010. Expect widespread use of ultra-low cost chips which use mains power sockets to talk to any other device in the house that has a plug. This allows plug-in-and-go low cost networking in every building anywhere in the world – as in my home. The brains of the system is any PC in the house, or a dedicated console which also controls the burglar alarm. With sensitive people-responsive temperature control in every room, light controls and other features, smart houses will boast 15 to 20 per cent energy savings. The Internet and telephone can already be transmitted down conventional power cables directly into homes, enabling high data flow using every part of the national grid. Every device with a power socket can be online. No more cable or telephone networks around the home. However expect many household name companies to suffer huge losses on 'intelligent home' technologies – few want such things as fridges with touch sensitive panels for e-mail and web on the doors.

Expect intelligence to be coupled with local background power generation in many homes from high efficiency solar cells, and from wind in rural areas. Expect growing numbers of homes in rural and urban developments to feature fashionable local sewage treatment, with methane gas recovery for heat and cooking, together with the recycling of 'grey water' from rain, bath water and washing machines, for use in the garden.

The other day I asked my eleven-year-old son if he would like to come shopping to the local large supermarket with me (he often does, and fills the trolley with food he likes).

'No way,' he said, 'but I'll do it on the Net.'

He's used to seeing our weekly groceries arrive in a van after an Internet order. Expect massive growth in online retailing.

No more shopping for food – even on Net

But we don't need to order at all: a small scanner costing a few pounds is near the rubbish bin in the kitchen. Every time we use a product up we throw it in the bin. If we wish, the scanner reads the bar code on the side of the packaging and automatically adds it to the shopping list for the weekly order. At the weekend several boxes arrive, containing all the items that have been used up – plus of course any other items added manually. Such a household never needs to shop again for groceries – unless it is for something special, say for dinner that night.

Expect Net shopping for regular items such as milk and bread, but going out shopping to develop into a major leisure past-time, an experience. Shopping malls have a big future in the digital world (see page 36).

Tomorrow's world will be dominated by owners of the biggest networks

Who is going to own the world's networks beyond 2005? Telecom requires huge economies of scale for survival and global domination. Therefore expect more rapid mergers and new alliances as large companies join in a frantic bid to carve up the world. It will be a fight to the death, with many smaller companies wiped out by speculative over-investment or fused into new corporations. Expect widespread concerns about monopoly power to be offset by rapidly falling telecom prices, and by the lack of any effective regulatory mechanism for global predators. Expect huge concerns about the costs of running 3G digital networks – with chaotic pricing policies – but 3G, 4G and 5G impact will be huge.

The Impact of Broadband/3G

What happens when old copper telephone wires or 3G mobile phones are always online and transmitting data at up to 2 megabits per second? A revolution. Broadband uptake/availability will be vital to a nation's future competitiveness.

- Book download   0.25 seconds
- 20,000 e-mails   120 seconds
- 2 hour videolink   7200 seconds

Two hour videoconference is equivalent to
- 12 million e-mails and attachments

Conclusions: Telecom companies charging for data flow, not connection time, will push the market towards video as the only way of maximising bandwidth use and revenue. Internet traffic will be dominated by video by 2010. Video phones will change how people live. Forget one to one calls. As reporting in the Iraq war showed, video is about entering someone elses world and experience, showing them what's going on.

Telecom merging

I no longer know whether I am using a phone, a TV, a computer or some hybrid of them all. It will soon be impossible to buy a PC without a telephone built in, or a TV without full Internet access. Every telephone will have a screen and a camera – not just mobiles. Already screen pay-phones with Net access are already widely available all over the world.

Internet replaces telephones?

The Internet offers users free phone and video calls worldwide. The quality may sometimes be poor, but what do you expect from the first 'days' of a new revolution? It is getting better. Our home PCs are connected on a local wireless network to the outside world using a telephone line which is fifty years old, but provides two megabits of bandwidth for continuous net access, with zero call charges. Today it costs $150 a month. Expect that to fall to less than $20. This ADSL system gives us unlimited global phone calls, and unlimited videoconferencing – all day every day.

So I can phone a relative or friend in Australia from the US and it costs me absolutely nothing. What is more, during the call we can both see the same documents on our screens and edit them together. If we both turn on our cameras we can see each other as we speak. The picture is good enough for me to do interviews on satellite. It is revolutionary, beyond the wildest dreams of most people even in the late 1980s. Life's too short to travel when you have virtual reality.

Keeping up is a nightmare for communications companies. Expect debt problems, takeovers and company failures. Overhauling BT's vast telephone network, replacing switches, changing copper wires to optic cable and rewriting software, will cost up to £27.2 billion ($46 billion). That's nothing compared to the costs of developing 3G mobile phone networks.

Expect all telecom companies to use Net-style tricks for voice calls by 2010, sending data in packets by different routes mid-call to get the maximum call capacity out of every millisecond with the minimum use of 'peak capacity networks' provided by competitors. Expect most electricity companies to use their own distribution grids as trunk lines for data transmission, with high voltage cables carrying heavy traffic.

A major challenge will be rapid shifts in consumer behaviour, as we saw with the dramatic growth of short text messaging on phones, which overtook voice calls, at the same time as total rejection by consumers of WAP web surfing on mobiles.

My mobile phone is a computer

For over six years now my only mobile phone has been a full Internet-compatible pocket PC with word processor, e-mail, fax, address book and everything else. It means I surf the Net on a train in a tunnel under Swiss mountains, and deal with my e-mail. It is programmable remotely, so my Net software was configured in a flash from a centre many miles away. Bought for $1,500 in late 1996, I predicted then that large corporations and financial institutions would soon be giving them away to their important clients, complete with proprietary software. Within a year the purchase price had fallen to $300. Phones will get larger (PDA size for multimedia) and smaller (earpiece size for voice calls only and speech activation). Expect all PDAs to charge smart cards, connect to banks, and to carry tiny cameras and colour screens, and handle voice calls.

Virtual retinal displays using low energy lasers sent from the phone to the eyeball could also provide full-vision displays as an overlay on everything else you see. That means you can walk down the street seeing a ghostly image of a road map in front of you. Everything connected to everything, everywhere, all the time.

Information apartheid

The world will divide into information haves and information have-nots. The privileged majority will accelerate ahead leaving a digital underclass far behind. But some in wealthy nations will totally reject the digital world – refusing mobile phones or e-mail as far as possible. Every nation will contain both communities and a nation's future prosperity will be determined by the proportion of haves and have-nots, something which will change fast in many poorer nations by 2010.

Keeping information secure will be a major headache with gnut-ella and similar software allowing unrestricted global access to billions of files on PCs. Napster's centralised system for sharing MD3 music files will spawn thousands of 'Peer to Peer' ways of doing similar things. Intellectual property and patents will be increasingly difficult to protect.

The end of books?
Non-fiction books work well on-screen, but reading fiction requires a long linear read and paper is likely to remain popular for decades to come.

Aesthetics will be important in publishing: the smell, feel, weight and binding of a 'real book', the quality of the paper and the kind of ink. Expect publishers to continue to integrate with other media companies to form complete information corporations. Expect the growing trade in multimedia reference books on CD-ROM and other electronic media to give way to online reference. Expect net competition for all non-fiction reference works, with users paying a fraction of a cent per page, and with photo printers in many homes and most offices. Expect digital books with more than a hundred double-sided paper-thin electronic pages. Just load the text you want, settle back and enjoy. But 'real' books will remain popular, partly because of the paranoia by publishers of digital theft.

THE INTERNET AND TV

I have access already to 50,000 TV channels
Meanwhile the Net is becoming TV. For example I can watch Fox TV (US cable) on my own PC at home in London down copper wires on the Net (free). My oldest son only watches webTV.

People rave about digital TV giving us 500 new channels, high definition, wide screen. But these need special boxes or decoders. I already have access to over 500,000 channels, which is the number of websites already broadcasting video on demand, mostly brief promotional clips. People say they have no need for that many channels, but that's because they have no vision. I don't watch much

TV – never have – but I would appreciate a Net with at least a million channels. That means that whatever I want to find out about I can read about, listen to or sit back and watch. It could be a page of text about new car models, with a video of the latest model being driven running quietly in the top right-hand corner. It could be a domestic version of the 'Big Brother' TV show.

I have launched my own TV station – on the Net. It takes less than five minutes to record a comment on today's news, convert it to a web TV site in a web page and add it to a server the other side of the world. My site gets up to 6 million hits every year. Right now Net TV for most people is poor quality – it takes the BBC 30 minutes to send a minute of full quality broadcast downline, but highly compressed smaller TV images can be watched live and broadband improves TV dramatically. Net TV will continue to attract advertising revenue. Hundreds of companies are developing Net cameras, microphones and headphones, for do-it-yourself TV.

Web radio is far more advanced

Thousands of new web radio stations were launched in the last three years alone. There is a vast choice. Capital Radio's Cafe in London is one. Their almost-live programme is reaching a new US audience. Expect tens of thousands of net 'radio' stations by 2010, including thousands of larger churches regularly broadcasting live services, and hundreds of academic institutions, parliaments and law courts, all pumping out a mixture of live and pre-recorded sound for those that click on their sites.

Expect tens of thousands of amateur disc jockeys, single-issue activists, eccentrics and misfits to be broadcasting to audiences of only a few tens to a few hundred from garages or bedrooms with virtually no equipment other than a hi-fi, a PC, modem and a microphone, maybe with TV camera or mobile videophone.

Mainstream TV companies lose audience

One thing is certain: mainstream TV companies are in for a severe hammering. By 1998 it was already the case that less than half of prime-time TV viewers in the US were watching ABC, CBS or

NBC. Many were wandering across to hundreds of small-time specialist cable channels but one in 20 prime-time viewers a year were just pressing the off switch. ABC lost 2 million viewers from September 1996 to May 1997.

By 1998 there were already over 200 'unrated' cable channels in the US with audiences too small and transitory to bear the cost of audience analysis. Traditional stations are fighting back with rehashes of sitcoms, game shows and repeat movies but it won't reverse the slide, which has continued through the first two years of the twenty-first century.

Television shows shaped by instant audience feedback

Live television shows will be greatly influenced by an entirely new broadcasting concept: audience scoping, or instant analysis of how viewers are feeling. Traditional audience shows rely on a studio audience to create feedback through laughter, rapt attention or signs of boredom. But studio audiences are completely atypical. They have chosen to be there so they are probably enthusiasts of the show, and they can't leave. In addition they have all the benefits of a live performance. They are completely unrepresentative of what is happening in living rooms across cities and the countryside. Expect live, interactive TV shows and reality TV to move from mainstream hype to semi-permanent niche broadcasts on smaller channels.

Expect TV companies to be able to see what is happening, using the Net and interactive TV. Every few seconds a dial will show them whether the last joke caused the audience to fall 2 per cent or rise 2 per cent. Fast audience falls are the result of instantaneous channel zapping and take place within two to three seconds of mild boredom or offence, while audiences soar if a percentage can be persuaded to hang on for a few minutes before zapping somewhere else or turning the TV off. Audiences are built by collecting and keeping viewers who zap into the programme, adding to faithful regulars. Desperate fights for audience will create bizarre behaviour when presenters can literally watch millions of people walk away or sit glued to their seats. Sensationalism is very effective in the short term (15 seconds to 15 minutes) but what will it do in the longer term? It won't arrest the overall slide. Great shows will build audiences in seconds as viewers alert friends by Text messaging. It's already happening.

Desperate TV companies pay people to watch

The situation is so desperate that ABC announced it would actually pay viewers to watch programmes, offering air miles in return for the completion of questionnaires proving they had watched certain programmes. While their advertising revenue is under threat the costs of quality TV productions continue to rise. The latest gimmick is to try to claim that although a station has few viewers, they are the right kind for your product. Individual channels in future won't have the audience size or cash to make high quality, high cost programmes.

Advertising will change from fast clips sent out to people who don't want them, to interactive adverts for people looking for information, coupled with ambient advertising – such as a product on a shelf in view of the camera during a show. Advertising breaks will be less effective as millions zap other channels instantly, or robots do advert cuts for them. TV ads will only work if they are entertaining or informative.

The TV generation is growing up and is getting bored. The old pattern of passive viewing is being replaced by much more inventive searching for what to watch. That's where Net TV will score highest of all: total choice. Instant access to 100,000 films starting right now at the push of a button. Three hundred thousand episodes of popular programmes available after a delay of less than 20 seconds, and all digital TV quality.

Expect TV, movies, music, words and software to be pirated at an unprecedented scale despite every effort of copyright holders. This will affect revenues and profits.

Expect TV clip watching on mobile phones to soar – whether the latest goal, a news bulletin or a greeting from friends. Expect news networks to use live instant news feeds from mobile video phone users who happen to be in the right place at the right time – and who get paid for their amateur journalism.

CYBER-MEDICINE

As a physician, cybermedicine fascinates me. Surgeons in California have carried out a series of minor operations using telemedicine: a 3D screen, metal control gloves and robots. Meanwhile in Britain specialists at Queen Charlotte's are using virtual ultrasound to diagnose tumours in people hundreds of miles away. However, the equipment is still expensive for operations, costing around £50,000 a time.

Telemedicine works

Telemedicine saves costs and lives. Why go to see a specialist when he can give you an instant interview in your own home? A recent study by Arthur D. Little estimated that telemedicine could save the US $36 billion a year if widely used, providing the following benefits as well as many others:

- home monitoring of high-risk babies after birth
- remote advice from experts to home or hospital
- immediate access to patient records in emergency admissions
- reduction in hospital stay
- lower ambulance costs

Virtually inside hearts and lungs

In the 1970s computerised axial tomography (CAT) was the biggest imaging breakthrough since the humble X-ray the best part of a century before. The huge machine had a revolving ring which turned around the head or body of the patient, taking a new picture every few seconds. What the doctor saw was identical to what she would have seen if an electric saw had sliced the person's body in half. But every slice took a few minutes and a full scan of a head took an hour. Very small children had to be sedated or restrained, as any movement blurred the slice.

Now next-generation machines take hundreds of images in the same time, rolling them together a stage further to build a virtual world of life inside the patient. A surgeon can 'travel' up a blood vessel, or down the windpipe right inside the lungs. What is more, he or she can practise complex operations using the 3D model.

Expect dramatic advances in medical imaging over the next five years. Because the machines are so expensive, and the output is computer data, expect a further stimulus to telemedicine as world-class specialists gather in global conference calls to decide on treatment in rare and difficult cases, and then bring in a remote surgeon, assisted by a local team, to oversee the procedure.

Let robots treat the sick

Medical students in future will need to know more about talking to people and less about treatments. What's the point in a head full of knowledge when 50 per cent of all medical knowledge is out of date every ten years? When care plans are created by robots, personal attention matters more than ever: touch, feel, empathy and understanding from real human beings, selected for their ability to understand.

Computer-assisted diagnosis has been used for some time. In 1979 I helped work on a simple diagnostic programme for headache, designed to separate tension from migraine, brain tumours and many other causes. After typing in a series of yes/no answers, the machine came back with a percentage risk for each possible diagnosis. It was primitive and inaccurate, but it was a start. But today we have the data to create such algorithms for a host of conditions and treatments.

It will become universal in some countries by 2010 to consult the computer before beginning a wide range of treatments, with doctors forced to use it not by law but by their insurance companies. If a patient dies from particular causes, the first question to be asked in a court will be whether the doctor consulted the computer system and followed its advice. The computer network will always be far more up to date and complete than any single doctor can hope to be, representing the collective experience of tens of thousands of physicians. Woe betide a doctor who insists in court that her judgement is superior to the computer when her patient is dead in unfortunate circumstances.

The first universal diagnostic systems are here already – on the

Internet. Capsule is a programme designed to advise doctors on prescribing. It was developed by the Imperial Cancer Research Fund and will soon be in use in British surgeries. It is an 'expert system' using a database of medical knowledge to make recommendations on all illnesses. In trials, doctors using the system were 70 per cent more accurate in their decisions.

The use of computer diagnostics and networks will never replace the clinician's judgement, and treatment will still be heavily influenced by the patient's own views. Computer diagnostics will mean that doctors can continue to practise safely for several decades after qualifying, with additional multimedia refresher courses, surgery videolinks and electronic performance tests.

Surgeons measured by the numbers their operations kill

Surgeons in the future will be measured by the numbers their operations kill and complication rates after operations, and perhaps by length of hospital stay. Expect a rethink. These crude measures encourage bad practice – for example early discharge, which may be safe but is stressful, painful and unpleasant. A doctor may refuse to operate on someone at risk of dying in surgery even though the patient wants to take that risk. Such life dilemmas as these will dominate medical practice throughout the next century – rationing, risk, individual and community benefit.

Within ten years we will have mastered the art of fusing microchips with living cells. Peter Cochrane, BT's former head of research, has named a new species: 'homo cyberneticus', a human being decked out with many different devices all working off body heat. Muscles driven by implanted computers will be common, together with other experiments for the more adventurous such as sight for the blind, using implanted cyber-brain techniques where electrical devices implanted inside the head are generating nerve patterns. Brain cells in rats have already been grown on the surface of chips. One person's eyes could become the camera for the person standing next to them. By 2015 expect the first cyber-brains in experimental animals, where digital data is directly accessed and stored. Cochlear implants are already giving hearing to the deaf, and chips implanted into the retina are already improving sight.

THE VIRTUAL CAMPUS – LONG-DISTANCE LEARNING

Virtual campuses are springing up everywhere. Thousands of people already attend college using the Net, with lectures, notes and advice online. Many executives are already refusing to attend on-site training, insisting on remote learning wherever possible. This is partly a reaction to 'learning overload' in a rapidly changing world where almost all their knowledge base is redundant in five years. In future students will pay for courses in modules by the number of hours online. Many campuses will provide lecture notes for free and charge only for 'live contact', or for e-mail replies or for marking course work. The Massachusetts Institute of Technology has already decided to make all reading material available online.

The technology is already here to provide 300,000 recorded lectures online, each able to start playing within 10 to 20 seconds of request. Then there are live TV broadcasts on the Net from lecture rooms. Your seminar could be visited by any one of half a billion people, with 25,000 or more participating and response to lecturers by e-mail or Internet Chat, voice links or even video links (a few at a time). Lecturers will find they have instant feedback, with questions tumbling in from all over the world.

Education never stops

The only way to cope with a changing world is to keep learning. Expect education and training to remain a core national activity at every level of society. Expect people to take several graduate or postgraduate courses in a lifetime. However, as a countertrend expect a backlash against paper qualifications, as employers realise that leadership, energy, dynamism, initiative and organisational skills are not created by studying books or going to lectures. Expect a rethink about the content of MBA courses, with far more emphasis on leadership and analysis, as well as on world-class communication and advocacy skills.

Expect governments to set increasingly ambitious targets for adult literacy, numeracy and computing skills, and for schools to be increasingly called to account for bad performance, not only by parents but also by past pupils. Expect many lawsuits by ex-pupils who feel their entire adult lives have been wrecked by poor teaching, failure to recognise special needs, failure to intervene in cases of bullying or sexual abuse or failure to stretch bright pupils to their full potential.

Expect a return to single-sex schools in many areas where co-education has resulted in tens of thousands of boys dropping out. Expect persuasive arguments that single-sex education for both sexes means sharper concentration and less distractions or showing off, especially with age of puberty falling to eight in girls (see page 93).

Expect a complete rethink about punishment and discipline, with the recognition that a no-touch policy isn't working. By 2005 there will be growing intolerance of a playground culture that takes threats, beatings and stabbings for granted, with strident calls for teachers to be able to teach without fear of attack from pupils or parents. Expect changes to come in steps, following particularly awful and well-publicised events such as the death of another teacher or the death of a pupil after savage bullying.

Expect tough new sanctions, including new freedom to suspend or expel pupils for anti-social behaviour. Expect growing expenditure on special needs schools for very disruptive pupils. Mainstream schools will not be able to risk keeping disruptive pupils as the emphasis grows on getting results.

Expect continued ghettoisation in schools, with people choosing a school and then working out where they need to live. Expect the final collapse of bussing and artificial attempts at black-white integration in US cities.

College libraries become fossils

What happens to college libraries? They start by computerising their indexes, but what is the real, useful shelf life of a book? Most science and current affairs books rapidly become out of date. Recently I was shown around a well-known US college physics department. All around the walls were thousands of scientific journals. I told the professor to throw them all away. Part of a dead world, of no interest except to students of the history of science.

Worthless and taking up valuable space. He agreed, rather sadly.

When scientific knowledge doubles every 10 years or less it raises big questions about learning. Most of what you learn is soon history, interesting but almost useless except in providing a general understanding. Universities still want people to learn facts, but the real money-making skills in the future will be searching and analysing databases.

Data location and rapid analysis are survival skills

In a world of information overload there are some basic skills that third millennial students will need. None of them is difficult to master. The greatest skill is text scanning, which is quite different from reading and requires no technology.

Faster Data Entry to Human Brain

- Sound via ears: 112 kbps
- Light via eyes: 3,000,000 kbps
- Lesson: Reading is faster than listening for data acquisition. One picture = 1,000 words, one video clip = 10,000 words. Phone calls are very slow compared to e-mails for information.

Most senior executives can scan 5000 words per minute. On the phone they understand 100 words per minute.

Text Scanning: A Third Millennial Skill

- Ability to scan a page of A4 text every five to 15 seconds and underline 90 per cent of relevant data with a marker
- Ability to scan an entire 250-page book and underline 50 key sentences in less than 20 minutes.
- Ability to write a two-paragraph, world-class executive summary on a company or product you have never heard of, using a wide variety of computer online databases, in less than 30 minutes.

In the past a top scientific adviser of a petrochemical giant might have received a phone call asking for a briefing on a new product just launched by a competitor. The adviser would have been expected to be able to provide such a brief from memory. Today's world is changing so fast that a brief will always be unsafe without an up-to-the-minute data search, and the product may be so new that many inside the launching company have never heard of it either. Human memory becomes irrelevant.

Large companies launch new products several times a week

The United Bank of Switzerland (UBS) is in the top super-league of banks worldwide. Although strongly Swiss in culture and top-level leadership, the bank is fully globalised, and is a highly complex operation with a strong retail presence in Switzerland in private banking, investment, wholesale, and a host of other dimensions, including new insurance industry partnerships. Almost every day new products are launched to particular client groups somewhere in the world.

Company knowledge is critical. Rapid communication, targeted to those who need to know, is the answer. The pre-merger old UBS was one of the first banks to introduce an integrated e-mail system capable of handling outside as well as internal mail in a secure and confidential way. E-mail was not the whole answer, but it helped. When first installed, the mail system gave them a firm edge over many competitors, an edge that of course rapidly disappeared without further aggressive investment in next generation technologies. Expect all the largest companies in the world to invest heavily in knowledge management systems. One of the greatest challenges in the twenty-first century will be to make these systems work in a way that wins competitive advantage.

Intranets are the key to universal corporate knowledge, allowing wide access to interdepartmental databases and hundreds of other internal information sources.

The key competitive challenge in the first years of the twenty-first century will be this: can the organisation ever learn, or does it stay ignorant as each person's knowledge remains inside that person's brain, a situation made worse by staff turnover? Loss of corporate memory was a key issue when downscaling was progressing at its fastest in the 1990s, and again in 2001/2. Knowledge management would have prevented some of that. Knowledge management will be a key survival issue in all large organisations over the next two decades, together with reducing turnover.

Training in future will be multi-dimensional

As a futurist, I lecture senior executives of the largest corporations in the world, board members, directors and others, on global trends. The medium is the message. Three huge screens bombard the senses with hundreds of images, sound, video, virtual reality, video conferences and animation. It all adds up to an image of the future. Technology takes us on a whirlwind global tour of continents, industries and issues, and blasts us in the face with inescapable realities, challenges to management for the next five to ten years.

The entire presentation is compressed onto an ultra-fast portable PC running film, three-dimensional imaging, Internet, video links. But that's the technology of today. Tomorrow's executives will expect loads more. Lecturers will be judged not just by content but by technology and the way they use it. People need to taste the future, reach out and touch it with their hands. 'In your face' experience is worth hundreds of hours of private study.

Expect a huge emphasis on multimedia in corporate presentations, with the routine use of rich multimedia formats in portable executive presentations to clients. Expect portable PCs with even larger flat screens, and good sound capability, designed to be good enough for viewing by up to 10 people in daylight without using a projector. Designing these new art forms will be a completely new industry, drawing on experience gained in making corporate videos.

Expect an anti-reaction by 'eccentrics' who make a deliberate point of not using any technology to present, relying entirely on person-to-person interaction.

Just dropping by – 5,000 miles away

I was setting up a three-hour workshop in Zurich for 30 executives from all over the world. It was 7.50 am and the first participants were gathering. My computer was busy throwing up onto the huge screen behind me a fast-moving mix of video, sound, and logical sequences – and was also sitting on the Net. I was just about to return to the first slide when I heard someone say 'Hello' behind me. I turned to see a smiling face up on the wall.

'Hi there,' he said.

'Hello,' I replied. 'Where did you just spring from?'

'Durban,' came the reply. 'I was just wandering about to see who was around.' By now a stack of others were gathering around his PC. My own golf-ball sized camera gave him a nice view of everything.

I explained that we were in Zurich and invited him to sit in on our workshop. He did, sitting unobserved as a distant participant for the first half an hour, before I flipped him back up on the screen to say hello to everyone and participate in discussions. His spontaneous arrival said more about the future than any degree of pre-planning. Of course he was able to call by only because I had told my computer to be polite and friendly to any cyber-callers – at least to answer the door so I could say hello myself.

It was the digital equivalent of a member of the public walking past the building and poking their head round the door to ask what all the high-tech displays were for. The only difference was that the Net created that same informal immediacy with someone more than 5,000 miles away. It just happened. That's the global village.

By 2005 mobile video phones will be everywhere you want and cost little more to use than an ordinary phone call today. You'll call someone and ask why they've turned their camera off. What have they got to hide? Holo-phones will be interesting when they come: producing three-dimensional images right in front of you of the person at the other end. The use of intelligent optical materials at either end will mean less computer power and bandwidth is needed.

VIRTUAL REALITY

VR creates hunger for more

Virtual reality will be a boom industry, especially in retailing. Expect VR to be used in shop windows to attract attention, inside the store to demonstrate products, and also in the home for interactive electronic ordering. Expect complete multimedia performances to pour out of supermarket trolleys in most large stores triggered by products you have selected in the past and your whereabouts in the store. Expect utility repeat shopping (bread, stamps) to be done from home and the rest to be entertainment-driven as a leisure experience, hence prompting a rethink about shopping centres that turns them almost into retail theme parks with wide age-range attractions.

Expect retailing to move towards, on the one hand, low-cost sales of products with minimal support, and on the other price-insensitive, highly personalised sales and support packages for people fed up with wasting half their lives trying to sort out problems with what they have just bought.

Expect the big shakeout of retailers to continue, with the loss of millions of smaller retailers by 2005, replaced by chains who will increase their own-brand sales from around 15–20 per cent in 2000 to 30 per cent by 2010. However, many corner shops will survive, particularly as car-use restrictions begin to bite, and as people become more ecology conscious.

Expect more products individually tailored to suit people's increasingly varied lifestyles. Benchmarking will continue to be popular but will lead to convergence and eventually competition on price alone. One example is in food retailing. Benchmarking of new services simply means all supermarket chains are likely to offer loyalty cards and financial services, but then where is the competitive advantage? Where is the customer value? Where is the real loyalty to be found? Expect a new emphasis on factors that will remain attractive to customers – not just price and corporate efficiency, both of which are easily copied and neither of which will produce lasting shareholder value.

Virtual reality will dominate leisure industry

Expect VR films and TV, with new tricks to convince each eye of the viewer into thinking it is receiving a different picture – without headsets. Expect virtual reality to dominate amusement arcades and theme parks by 2010 with ever larger worlds, body suits, headsets and vast virtual reality cinemas.

VR in manufacturing

Expect all prototyping to be modelled in virtual reality by 2005. Expect many more examples like Land Rover, who detected several serious flaws in their plans for a new production process through VR. VR tells you if an engine can be fitted in a space, or whether access will be possible to a gearbox for a machine.

Expect huge advances in rapid prototyping with routine component production from VR images. Computers will drive lasers to carve solid structures using a selective layer sintering. These can then be assembled as parts of a machine, toy or vehicle before casting in plastic, ceramic, metal or other materials.

Primitive VR worlds already exist

The other day I walked in cyberspace down a street and into an art gallery. There I met the person who built it and talked to him about his art. He could see me and I could see him. What was unnerving was that we were in different countries, and so were all the others who joined in our conversation. I have also walked into a branch of a virtual bank and watched someone talk to a teller and take out a loan. Each of us had a computer-generated body which could walk, bow, wave, jump or whatever. In future these three-dimensional images will be created directly from the TV cameras attached to our PCs. The only technology we needed was a PC and an ordinary telephone connection – and the software was all free.

Expect people to enter global online VR computer game worlds with three-dimensional characters based on photographs of their 'real' selves. I can tell you it feels a little strange.

Nations, regions, communities and cities will rush to become digital societies. These will be areas or cities where governments have committed themselves to massive investment not in electricity, water or roads, but in community electronics. Some day I am going to stake out a piece of land in a cyber-world and build myself a house and an office, and a lecture area perhaps, as well as a meeting place.

Reaction against robot contact

Expect a reaction against primitive human-replacement technologies such as telephone voice response units (VRUs). Let's be clear about this. When contacting a large organisation with many departments and possibilities there is no substitute for being answered by a knowledgeable and efficient human being. Listening to a 60-second list of options, and then another list and then another, may save the company time but my time costs me money.

It certainly saves them a fortune. The cost for a banking transaction by phone is $1.75 compared to $4 at a branch, according to Mentis Corporation. However, the cost falls to just 25 cents per transaction when the whole deal is done by VRU. If I wanted to make a donation to the operating costs of the company switchboard that would be fine, but I don't. I rang wanting a service and here I am doing them a favour. What happened to customer support?

Speech recognition faster than we can think

Part of this book has been dictated not written, and my portable computer has converted my continuous speech into reasonably accurate text at up to 140 words a minute, using a vocabulary of 200,000 words. That speed is equivalent to up to 8,000 words an hour, or writing an entire book in 12 to 25 hours – faster than any author can think.

A little slow, perhaps, for those wanting 160 words a minute of continuous speech, and only 90 to 95 per cent accurate, but then how accurate are you typing 3,000 words an hour, eight hours a day? And this is only the first day of a new speech recognition age. The medical and legal costs of repetitive strain injury could mean insurers will insist that staff speak, not type.

Expect every major chip manufacturer to rush ahead with dedicated speech recognition chips which then become built-in features to a vast range of products, ranging from washing machines to vending machines. The best products, hardware or software, will win a big slice of a multi-billion dollar market beyond 2005, with companies battling over 1 per cent or 0.5 per cent improvements in accuracy at speed. There is a huge difference between having to correct every tenth word and only one word in 50.

Voices take the strain

And then will come repetitive voice injury as people find the discipline of clear, accurate speech strains the larynx. But this is today's technology. Within ten years it will be standard practice to speak rather than type for many who never bother to master a keyboard. Have a look around your office – even with 2003 software, anyone you see who is typing large amounts of text is probably costing you money. Expect the economics to shift firmly in favour of speech recognition by 2006, with pressures to redesign open office areas as a result. Speech often needs privacy.

ARTIFICIAL INTELLIGENCE

Some artificial intelligence purists say that true artificial intelligence does not and probably will never exist. They are both right and wrong. Right in that at present we don't understand how to make computers think for themselves in a sophisticated way, wrong in that humans can be tricked into thinking and feeling that they can.

In 1978 I was involved in artificial intelligence research based at the National Physical Laboratory in West London. Our aim was to develop systems for interviewing patients at Charing Cross Hospital in London that would be indistinguishable from the doctor. In one famous experiment on an earlier system a patient was linked to a psychiatrist via a computer and screen in the next room. They communicated by typing to each other. Then the link was cut and replaced by the computer itself. The patient was unable to tell when the person listening and responding so intelligently was a computer and when it was a human being. Artificial intelligence has come a long way since then.

Male and female robots

Expect to see dramatic improvements which will affect everything from continuous speech recognition (understanding of context and grammar) to search agents for the Internet.

Expect to hear male and female robots at the end of the phone, and the same reaction against them as for primitive Voice Recognition Units today. In a fractured, chaotic and fast-moving world people will willingly pay a premium for human contact. For a single person living on their own and teleworking at home, a genuine human being on the other end of the line may be the only personal contact of any kind that day.

Expect premium pricing when you want to touch another person's life, when you want to know that another human being will take some action as a result of what you have just said. We are already seeing such price differentials in banking, with some banks charging customers a fine every time they use a human being face to face to deposit or withdraw money.

Handwriting recognition

So what is the point of learning how to type? Some people who can't stand using keyboards swear by their portable handwriting units, which convert words written with a stylus into text. The trouble is that writing will always be slower than fast typing, and less accurate. Why bother?

Expect to see hundreds more 'handwriting to text' products for a generation who still like to scribble. Expect them to fall out of favour within 10 years, swept away by speech recognition and other new technologies.

Typing skills will always be important

Typing skills will continue to matter – not least of all because dictating of any kind requires a reasonably quiet area, one where no one else will be disturbed or distracted, and one where sensitive issues will not be overheard. Note-taking in meetings will also continue to require a completely silent and unobtrusive recording method. Anyway, who wants a complete, verbatim computer-generated transcript of every word spoken, outside of Parliament or a courtroom?

Expect speech technologies to reach the point by 2030 where they can read speech silently, by lip and mouth reading, and monitoring movement of the larynx.

NET THREAT TO THE FINANCE INDUSTRY

The finance industry is being massively impacted by the speed of technological change. Every one of us uses money every day. That's why the financial services revolution will have a profound impact on all of our lives.

The Internet could destroy traditional banking

Banks will be little more than wholesalers of financial services after the big post-millennial shakeout. Their retail business is already being eaten away by food retailers, insurance companies, clothing stores – anything and anyone willing to make the move into selling credit cards, loans, mortgages, current and deposit accounts.

In the UK Sainsbury's food retailers decided to open almost 300 banks, while Tesco's and others were close behind. Every high street chain you can think of is ready to gobble up financial services market share. Pensions with potatoes, mortgages with milk.

Banks made the mistake of listening to their customers. Banks said: hardly anyone wants Internet technology and it serves no real market need. Then many banks began to panic as they saw a new cyber-world being built right under their noses. History suggests that few conventional banks will be able to accelerate fast enough into the cyber-world to be able to make a big impact. Despite the dot com crash in 2000/1, use of the net for financial transactions has rocketed. History will record their death by attrition, merger, and acquisition followed by cannibalisation of the business.

Banks call the new challenge by many names, such as cyber-banking – that is, they want to transfer traditional banking onto a new distribution channel. But people won't even want to use money in the same way, or need it in the same way.

Banking as it is will never survive

Network computing will destroy the power base of big traditional banks. Some traditional banking services will survive – at a premium, for those prepared to pay or with no access to technology. And many consumers will stick to old ways, for a while longer.

But banking as it is will never survive. Banks made their profits by collecting and hoarding cash, and lending at interest. But when cash itself ceases to exist, what then? In an electronic society there is nothing physical to collect or give out. Banking becomes a meaningless concept, used to describe a defunct industry which trades not cash, but electronic impulses.

Internet Banking Facts (US)

- More than 5,000 banks, thrift and credit unions had websites by 2000
- 1,556 fully transactional by 2001
- 44 per cent had poorly developed online security/risk management
- 10 million were banking online by 2000 expected to rise to over 40 million by 2005

Virtual money for a virtual world

Money becomes bits of data in a virtual, electronic age. But who needs a bank for that? Strongholds and safes disappear, as do security grilles and guarded delivery vans. Trades happen on any computer. The bartering of bits can happen anywhere on the face of the earth, at any time.

Any product, any channel

Forget about banks. In future you will buy any product, via any channel from just about any source you could imagine.

For individual customers it will mean instant account access anywhere in the world, complete automatic updates to their own home-based accounting packages, instant decisions on loans and mortgages, and better interest rates. Expect many Net banks to pass on lower costs as added customer benefits. Partnerships and alliances will be the key.

Online Banking Services

- Bank branches – pay for the privilege of dealing with humans
- ATMs – cash/electronic card machines
- Buy by phone
- Use kiosks with video links to experts
- PC banking with Quicken/Microsoft Money and others
- Internet banking and shopping
- Commercial online service banking
- Interactive TV

Most senior bankers I talk to recognise it is only a matter of time before a mortgage will be switched from one lender to another by an authorised broker, at a single mouse click. Once the legality is established, expect such loans to be moved up to several times a day by broker robots constantly looking for better deals on identical terms. It will spell death to traditional lender–borrower relationships.

When the history disks are written . . .

At the end of the twenty-first century, when the history disks are written, they will record that 1997 was the year of the Internet bank. At the start of the year few banks offered transactions on the Net, but by early 1998 over 60 per cent of European banks and a similar proportion elsewhere had recreated themselves as virtual banks in cyberspace. They were driven by fear over what their competitors were up to, and by the elusive dream that Internet transactions could cost them only 1 per cent of the cost of those conducted in the usual way. Unfortunately for them, there were no real savings to be had without making staff redundant and closing branches. But that takes time, and branches are hard to dispose of.

Four Fatal e-Business Errors

■ 1996 'The net is irrelevant – spend nothing.'
■ 1999 'We're late! – Spend everything!'
■ 2001 'The net is unimportant – spend as little as possible.'
■ 2003–5 'The net is just a successful distribution channel.'

Each error caught companies off-balance and led directly to the next over-reaction. The next error will be to fail to realise the Net's power to totally transform every part of the supply chain and internal organisation. Too many still see the Net as websites and sales brochures.

Regulation will be difficult. As things stand any bank in the world can allow accounts to be opened and run by someone in the US without any US controls, so long as deposits are not taken in America. Cash can be given out using credit or cash cards in machines, but deposits must be posted or wired abroad. In practice there are more controls, but only because few international banks will risk being unpopular with the US government. Expect scams and frauds. One offshore bank has already appeared and disappeared again after taking substantial deposits.

Online investing boom

Stock-brokerage will never be the same. By early 1997 there were already 1.5 million investors trading their own stocks on the Net, a number increasing by over 100,000 a month. Commissions plummeted, threatening traditional stock-brokerage. One discount broker, E*Trade, had $5.5 billion managed and 182,000 active accounts by mid-1997. The dotcom crash did not wind back the clock. I have used my data phone to obtain a quote on selling a million shares at $7 each. The commissions from traditional brokers would have been anything up to $3,500 but my data phone found me $20 fixed-price bargains.

So what happens to brokers? The fact is that brokers are in for big trouble, but only a few have woken up. The big stock markets in New York, London and Tokyo are in flat denial. If advice is free and you pay for the deal, what is to stop me phoning my broker for advice and then doing the deal on the Net? How long will it take for my broker to realise? I was talking to a group of private bankers recently. I told them to start analysing phone calls and share volumes. Many people will deal with brokers just to keep the relationship sweet – a deal here, another there – while switching more into computer trades.

Commissions are now so low that people are entering and leaving the market during coffee breaks – and making a profit. Even the smallest movement in share prices is worth speculating on when deals are basically free. Once deal prices have dropped to $12 or less as a fixed price, the amount becomes too small to be worth the bother of collecting for large banks. Do the deal for the customer as a completely cost-free loss leader, enticing her to come back for other services for which you can charge, such as comprehensive portfolio management or business loans. Some bankers sat back and relaxed after the dot-com crash in 2000/2001, deluded that somehow the threat would melt away – but every day it is growing.

Death of the national Stock Exchanges

The days of national Stock Exchanges are coming to an end. Companies don't like them because their global business is greater than a single exchange. Investors don't want them – they want to trade online 24 hours a day. Technology doesn't need them – because a single server in one building can handle all the mouse clicks. Expect dozens of new 24 hour virtual markets of which one or two will achieve rapid global influence. In the meantime expect thirty traditional exchanges in Europe to react with various degrees of apathy or alarm, and with a range of new alliances and partnerships. Expect no more than 10 survivors in 10 years. Virtual exchanges will cut dealing costs by more than 85 per cent, encouraging risk-taking, growth of volumes and huge liquidity – which will be the most important success factor.

Death of Stock Exchanges – online dealing

- $12 for unlimited share volume
- Small investors get current price data on-screen
- Buy/sell at work during coffee breaks
- Zero commission deals will be common
- Make your own cyber-market
- Companies start selling their own stock online
- National stock markets begin to die
- One global Stock Exchange – virtual

Traditional brokers have their heads in the sand. They imagine that the 'trust' factor will save them, that people will be too scared to do big deals on a PC. That may be true in 2003, but it won't be in the future. Anyway, when commissions are low or free a customer can spread the risk across a large number of smaller trades.

The fact is that technology means investors are not going to let themselves be charged thousands of dollars a year to have some broker click a button for them on a PC in an office elsewhere. Some argue that busy, wealthy CEOs just don't have the time or energy to fiddle around on their PCs buying and selling stocks. That's true for many, but they all have secretaries, junior staff or teenage children to do it for them. So the future for brokers lies in what they are best at: giving expert advice, making sense of all the data. But will customers be prepared to pay the necessary price for a phone call? Few will pay for a call what they would have previously paid as a commission on a fat deal. Expect many mid-range and smaller brokerage firms to slash overhead costs and struggle to diversify.

Wholesale finance, risk management, every area will be affected to a greater or lesser extent by cyberspace. Even those clients that insist on doing all their business in conventional ways will expect instant online access to reports and analysis. And pressure is growing on performance – it is rare for an investment corporation to consistently outperform market tracking funds, after deducting management charges. It all depends on who is in the team – and that keeps changing as people move to competitors.

Service is the name of the game

As in every other industry, service is the name of the game. There can be no other route to long-term success, unless you are competing on price alone, and that will be increasingly difficult. When people stop using bankers to bank or traders to trade, what will they want? When the average executive in a large company has instant access to world data in breathtaking volumes, you can guarantee almost by definition that he'll be more up to date on what he wants to talk about than his adviser.

So value-added service is a comprehensive, well worked out world-view, placing every new event into a global frame that makes sense – to everyone. That's the future. Clients overwhelmed by the background noise of data flows want value-information not just data. The more the data, the more the confusion.

Heavy techno-investment

No post-millennial survivor is going to make it in good shape without heavy ongoing techno-investment. The trouble is that most boards have either no IT expertise or people so senior that they are out of touch and out of date. How can a company possibly hope to stay competitive if such people are their main source of vision about new technology? So part of techno-investment is buying in techno-vision: high-impact communicators who have long-distance vision; people who can help us see beyond the tools we have today.

The technology is changing so fast that just planning with today's tools and next year's new models, or even those of the year after, will leave companies badly lagging.

Those who are often right will make a fortune

Trend hunting in the future will be a far cry from the seventies or eighties, when everything was more certain. In a globalised market there are too many variables for back-projection and forward-projection to work reliably. The people who saw cyberspace before it was born were radical visionaries, boffins playing around with bad modems, crackly lines and missed connections, eccentric hacks who dreamed of the impossible because it would be fun. They were people who weren't burdened by a bottom-line profit.

That's why economists don't make good futurologists when it comes to new technologies, and why so many boards of large corporations are in such a mess when it comes to quantum leaps in thinking.

Second millennial thinking will never get us there. It will just produce better versions of second millennial products for a dying species of second millennial elderly people. Expect some corporations to unexpectedly throw up their hands in despair and give up the third millennial race altogether. The logic will be that they decide to accept the inevitability of a dying business, with volumes declining, but remaining profitable as long as possible with premium products and niche marketing. Some institutions would prefer to die than to change.

A senior board member of a Fortune 1000 company told me recently: 'I'm glad I'm retiring so I don't have to face these decisions.' The trouble is that he hasn't retired yet, hasn't taken the decisions and is preventing others from doing so.

'What can we do?' another senior executive declares. 'We know our industry is basically dying.'

Expect plenty more fatalistic comments: whole boards that try to convince shareholders that the greatest profitability will be to scale down and down and then sell out, rather than take massive risks into unknown areas with fierce competition.

More options kept open

Parallel tracking spreads the risk. Take an insurance company which to date has been relaxed and sceptical, watching competitors 'waste a lot of money' building web sales. If that company changes its mind tomorrow morning and decides that, after all, it does need to start using the Net as a major sales vehicle, the trouble is that from a cold start it will probably take that company anything up to two years to catch up with where its competitors are today.

A smart parallel-tracking company has set up a site and is running it as a relatively low cost pilot, with many features and innovations and a wide range of functionality including ordering and staff interaction, perhaps even video links and international voice calls online.

Internet security

Security on the Net is a boom industry with hundreds of products and companies. Net fraud cost $1.6bn in 2000, expected to be $15–30bn by 2010. Security is a vital issue for every individual and corporation using the Net. Can people intercept my e-mail? Are my credit cards details safe? However, there are just three questions when it comes to online security:

- Confidentiality – is the line secure from listeners?
- Authentication – is the computer really part of the bank? Is the other computer really owned by the customer?
- Identification – is the user of the customer's computer really the customer?

We also need to ask how secure other systems are before demanding an overly expensive and inconvenient solution for online systems. Systems available today offer security many factors greater than that obtained with normal telephone instructions, fax or e-mail. A US survey shows that most people are more worried about personal details becoming public than about actual fraud. The fact is that every month, millions more people take the plunge by making their first purchase online, or by transmitting highly confidential information by e-mail. This is a channel we can trust, so long as it is used wisely, for example by encrypting e-mails where necessary, using technology which is almost unbreakable except by government supercomputers over weeks or months.

BIG BROTHER IS WATCHING

George Orwell's book *1984* gives a chilling picture of how technology could be used by a dictator to control millions. But the tools available today are already far advanced beyond what Orwell saw.

Using today's technology it would be cheap and practical to place a tiny camera and microphone in every room of every home wired to the Net, that could be dialled up and interrogated by Big Brother. It would add a negligible amount to the cost of any new building.

Privacy died a long time ago. I recently demonstrated the latest bugging devices to some company executives because they were ignorant and naively vulnerable to attack. High quality colour cameras hidden in the head of a tiny Philips screw, pens transmitting perfect sound a third of a mile. Little bugs hidden in plugs that can be listened to on the Internet the other side of the world. Boxes to decode even digital mobile phone calls – said to be secure.

During a lecture, I bugged one of the participants without him even realising. I then gave the receiver to all the others, one by one. The sound was so brilliant that they failed even to work out who was carrying the bug. They were horrified when I told them the bug was so undetectable that it would be carried by that person right back to his hotel room, and that sitting here, a third of a mile away, I would be able to hear every word he said – even know what television programmes he was watching. He started looking in his case but he was walking around with it on his own person. He was completely unable to find it, even with help.

Relatively cheaply you can buy a laser bug, which can pick up conversations inside an office from up to half a mile away, by bouncing laser light off the window. As the window vibrates it alters the light signals received.

So you are going to hear a lot more about privacy laws, surveillance, encryption and other related things. But laws will not prevent invasion by people intent on taking your company's best secrets. Market-sensitive information is worth tens of millions of dollars. Just one phrase overheard is enough.

Bugs can be hard to detect with machines. They can be turned on or off by radio signals from a listening post on the other side of the world. Five minutes after a meeting is due to start the bug is turned on – for a few seconds. If the room is quiet, the bug is killed again. If the listeners strike while a meeting is in progress then they begin to listen. When the meeting is winding up the bug is put back to sleep. A sweep of the room before or after the meeting will never detect a transmitter of such kind without very expensive and sophisticated room analysis, looking for frequency echoes that might be coming from a microphone. The sweep might take the best part of a day, and in theory needs to be repeated every time the room is used. Most bugs are placed by staff, so the vetting of visitors is not enough. Anyway, vetting visitors is impossible without stripping them naked – or worse.

Your own staff are the biggest security risk

Few board members have caught up with the implications of all this. They naively rely on security guards and electronic door keys. Cleaners, secretaries, junior and senior staff can be tempted by very large sums of money. And you will never know how often it happens because your competitors are hardly likely to tell you how many times they have been stung, even assuming that they know themselves. When it comes to markets, share price swings can alert suspicions, but other intelligence is far harder to track.

Expect new, aggressive anti-espionage measures, another name for spying on your own staff. Bugged offices, hidden cameras, secretly recorded phone calls, intercepted mail, duplicated e-mail and fax messages, networks which routinely explore every last byte on every hard disk. The bugging of staff in their own homes, and clandestine surveillance, will raise the hackles of civil liberty groups, and many other ethical dilemmas will be thrown up when surveillance teams uncover more than they bargained for, say evidence of other serious crime or links with terrorism.

In many countries it is not an offence to listen in secretly to a conversation on your own telephone or in your own room. Therefore the bugging of a company office with authorisation from those running the company is perfectly within the law. Bugging of staff in their own homes is illegal. Recording telephone conversations of telesales teams, however, is a completely accepted practice 'for quality control'. Expect this to extend to acceptance that a company paying a worker has every right to observe the person at work to monitor performance, whether that person is aware of it or not.

In 1996 Congress finally made economic espionage a criminal offence, when a corporation or institution steals information from another. With the end of the Cold War, foreign spy agencies are being increasingly turned to economic spying, including (in the past) the CIA. The CIA itself says around 20 countries are trying to steal each other's technology, or industrial secrets from the US.

Biolock security

In the future, banks, airlines, immigration officers or dictators will know exactly who you are from the pattern of blood vessels in the back of your eye as you stare at a screen – as distinctive as a finger-print (fingerprints themselves may also be scanned). Tiny iris move-ments prove you are alive, not a high resolution photo. Sure, there will always be industries of people offering ways to cheat these devices. Expect security measures and attacks to continue a roller-coaster race, with claim and counter-claim by industry experts and packs of anarchistic hackers.

Expect a whole new super-breed of techno-freaks whose entire waking existence is devoted to busting electronic security systems. Some will be terrorists, others just bored ex-students, but others will be earning huge sums from large companies who have recruited them to wage war against their own systems as a means of testing them. Others will be self-employed, in a shadowy world where their discoveries are sold for easy money.

Bank A is rung by anarchist B, who says he can prove that $45,000 was transferred from one person's account to another the previous night and back again. He says he will tell Reuters press agency about it if he isn't rewarded for his efforts in 'testing the bank's security' for them (for free). The bank pays up, rather than risk the embarrassment.

Very few people are ever prosecuted for breaches of bank secur-ity. In most cases the culprit is a member of staff. He or she is then asked to leave, and is given a wonderful reference to get a job with another bank. Everything is kept quiet. The police are not informed. Six months later history repeats itself. This pattern is common.

As I say whenever I talk about the future, people talk to me about values and concerns they have. Where will it all lead? What kind of world are we creating? We will find the answers in the final chapter.

OTHER CONSEQUENCES OF THE NET SOCIETY

Internet addiction

So what is real? Cyber-reality is just as real as life on the phone, if it is interaction with a group of people in their own identities in different parts of the world. But what about assumed identities? Can people be so caught up in cyberspace that terrestrial life all but ceases to have meaning for them? The answer is yes, in a minority, according to a recent American Psychological Association conference on Net addiction.

Cyber-addiction is becoming a recognised medical problem: the Net is the most important thing in life. Wanting to be constantly online, getting a buzz from use, needing to spend more and more time on it, withdrawal symptoms, conflicts with work and family. A new support group has been launched to help, called 'Caught in the Net'.

But then TV soaps are addictive too, and have been cited as a major cause of people moving away from eating meals together. They are accused of encouraging a nation of 'grazers and snackers', millions eating in front of the television in silence.

Technology driving us all

The cars of tomorrow will be computer-driven and steered, with automatic speed regulators, road sensors, radar and cameras to detect vehicles in front and behind. Cars will be fully networked, allowing weather warnings derived from the activation of windscreen wipers in cars ahead, or pre-paid tolls, booking of hotel rooms, video links to home and office, and automatic journey progress data to others who need to know where you are and when you will arrive.

We will see road trains, large numbers of cars linked with electronic towbars. Studies show that traffic volumes and speeds can be increased. Those who step out of line could be identified electronically and fined before they even reach home. But hackers could cause havoc. A single keystroke could throw a national motorway network into chaos.

The big debate will be over insurance cover and acceptability. The technology for all this already exists, but car manufacturers fear just one motorway pile-up could have the same effect as a fire in the Channel Tunnel, with the subsequent loss of confidence and damaged sales.

Bio-chips and bionic people

Salmon are already being labelled with chips implanted under their skin, and the same is happening with domestic pets. The fish devices transmit their identity to monitoring stations as they travel up and down rivers. The chips require no internal power source, are activated by external magnetic fields and last forever.

Expect the widespread tagging of human beings for identity and control purposes by 2010. The first step is already here, with Swatch watches that tell machines who you are as you approach. I have a watch which allows me to walk up to a computer and be greeted by name, and then be logged automatically onto my e-mail account. The next step is only to take the chip out of the watch and slip it under my skin. Activated by magnetic fields around me, it will require no battery.

Expect probation teams and other 'control' groups to utilise this technology for the electronic surveillance of people's movements, as an agreed alternative to imprisonment.

Biology could drive the computers of tomorrow. The Laboratory of Computer Science at Massachusetts Institute of Technology (MIT) is working on DNA computers, taking advantage of the incredible miniaturisation in every cell, where a blueprint for an entire human can be packed in a nucleus. Twenty to thirty years away perhaps – possibly less.

The artificial intelligence of tomorrow will be self-teaching robots, able to form new thoughts and make new suggestions, able to think creatively – at least they will appear to be doing so to those who interact with them. MIT hope that Cyc, a new creation, will be able to run its own research lab by 2020, designing unique experiments to uncover new knowledge.

Sony believes that one day we will be carrying little pocket devices with smiling faces that we can talk to and interact with, people we create to keep us company and in touch. In the meantime, millions of children have already been brought up with pocket-sized virtual pets which clamour (with a bleep) for attention during the night or at school, needing to be 'fed', 'watered', 'groomed' and 'exercised' to prevent early death.

The BT network is already developing its own intelligence, constantly sending out electronic 'ants' to test the speed of different routes, and switching traffic almost instantaneously in ways that used to take several minutes.

Informatics

We will hear a lot more about Informatics, combining computer science, artificial intelligence and cognitive science – building robots to clean hospital corridors, mow lawns, clean carpets or farm land. Efforts so far have been primitive. The hospital-cleaning robot at Northwick Park in Harrow, England, had to be retired after problems with people and other obstacles.

Another example is text analysers which condense a 50,000 word document into a two-paragraph executive summary – automatically – and presents it in perfect English. They work by analysing common phrases and ideas.

So then, wherever we look the world is getting faster, driven by technology and communication. We need a practical strategy to survive the challenges to management – and to survive as individuals.

CHALLENGES TO MANAGEMENT

How flexible is your organisation?
◆ Can your organisation adapt and change fast enough to survive?
◆ What strategies do you have to accelerate the adoption of change?
◆ Are you using training, seminars, workshops and conferences enough to set in place the new culture across the entire organisation?
◆ Are you using enough outside visionaries and motivators, world-

class communicators able to say some of the tough things that are harder to say for internal people?

Parallel planning

◆ Since planning takes longer than events to happen, have you invested enough time in parallel planning?

◆ Do you know what steps you will take, and by when, if a situation changes rapidly?

◆ Will those outcomes be in place fast enough or is more parallel investment needed now?

Telecommunications

◆ Are you getting the best value for money in telecommunications?

◆ Given the price wars, with the situation changing by the month, and the ease of transferring the use of the same telephone numbers from cable company to cable company, when was your bulk buying last reviewed?

◆ What is the corporate policy on mobiles?

◆ What is company policy on satellite phones?

◆ Have you considered issuing all staff and important clients with free pocket mobile video phones which are also computers?

◆ Have you considered placing telecoms and air travel under the same 'communications' budget heading, to encourage conversion of expensive air travel into high tech cyberlinks and other communication technologies?

Call-centre subcontracting

◆ How much of your incoming telephone traffic could be handled by a sub-contract to a highly skilled dedicated call centre?

Cross-selling

◆ Are you properly geared up for intelligent cross-selling, combining product data with customer profiles, available at the time of every contact?

Internet access and use

◆ Do all your staff have access to an efficient secure e-mail system, internally and externally? Do they all have Net access?

◆ Do you openly monitor activity, and site visits, in order to discourage abuse?

◆ Are you taking advantage of free or low cost online data such as free real-time share prices?

Website value

◆ Are you as proud of your website (your virtual HQ) as of your real HQ?

◆ Does it promote the company effectively?

◆ How does it benchmark against your immediate competition?

◆ What does it need to steal a significant lead on competitors?

◆ What is your five-year Internet strategy and are enough options being kept open?

◆ Have you properly explored new partnerships and alliances to keep in the mainstream and block others out?

◆ Do decision makers fully appreciate the urgency of how quickly this global market is being sewn up?

◆ Are people ever going to find your website once Microsoft and others start developing the channel with advanced technologies?

◆ How are you promoting your site?

◆ How are you monitoring site activity?

◆ Do you know which pages are most popular and why, where people arrive and where they get bored or fed up?

Encryption and security

◆ Are you using long enough encryption keys?

◆ Are you using encryption everywhere you should?

Futurology

◆ Who are the visionaries and motivators in the company who can give direction and purpose to the revolution?

◆ Do they have enough profile and platform?

◆ Who is looking beyond mere improvements in today's technology to help prepare you for next-generation technologies on which the future of your whole operation may depend?

Corporate knowledge management

◆ How is your whole organisation's intellectual capital and knowledge base being managed?
◆ Are you harnessing intranet power adequately to keep ahead?

Avoiding overload

◆ Who is summarising trends and data?
◆ Do those people understand your priorities?
◆ Do you have one key publication or summary that keeps you
   abreast of most important changes?
◆ Are your key staff trained in skim-reading/rapid scanning?
◆ Are you keeping up to speed with new technology – do you need
   help?

Multimedia investment

◆ Is your company measuring up well in a multimedia age –
   whether at shareholder meetings or in one-to-one presentations?
◆ Do you have the right technology in the right places, e.g. sufficiently high resolution data projectors with adequate sound
   capability?
◆ Do people know how to create world-class multimedia presentations?
◆ What support do you have in-house?

Virtual reality

◆ Have you considered the potential of virtual reality – product
   development, product promotion?

Reaction against robots

◆ Have you prepared for a growing reaction against automatic
   switchboard systems with push buttons or voice triggered selec-
   tions, and against voice mail?

Speech recognition

◆ Continuous speech recognition is one of the most important
   office tools ever to emerge. Who is evaluating such systems for
   your company, and how will your company change its methods
   of working as a result?

Financial services revolution

◆ Is your company ready for a complete paradigm shift in every area to do with finance and financial management?

Bugging and surveillance

◆ When did you last have a corporate health check regarding commercial espionage?

◆ How vulnerable is your business to a competitor listening in to board meetings or other conversations?

◆ Does your senior team realise how easily their own confidentiality can be compromised?

PERSONAL CHALLENGES

How do you cope with constant rapid change?

◆ If you find continuous change stressful, think now about creating areas of your life that will hardly change, and invest in them. Then you will find the areas that change most are less of a problem. Friends, family, a collection of treasured books, your garden, a regular round of golf – make your own slow-changing area. The greater your personal stability, the faster you can integrate change without stress overload.

Tracking trends

◆ How do you track early trends?

◆ What newspapers do you read?

◆ Are you scanning the content of your 'trade magazines'? They will often help with advance warning of change.

◆ Have you signed up for hot news – e-mailed every day or hour, just containing items matching your interests? In a fast-changing world, getting vital news half a day ahead of the rest can mean the difference between success and just surviving.

Keep learning

◆ When did you last learn something new and unrelated to what you 'do'? Stretch your brain, keep fresh, take mental exercise, be interesting. The broader your horizons, the greater your vision of the future.

◆ Are you spending time with people who stimulate you to think more widely?

◆ When did you last assign time to think laterally, out of the box, outside the office, with people outside your own discipline, and area of work?

Make your computers work harder

◆ Do computers scare you or get you excited? This is a make-or-break area for future success, so wherever you are on the scale, get in deeper. Learn from others, rather than battling on your own. Make sure your own computer knowledge is at least doubling every year, to catch up with accelerating PC power.

◆ Can you touch-type? Keyboard use will be important for the next ten years, so get going faster with a PC-driven typing tutor.

◆ Are you using the latest speech recognition systems?

◆ Have you given yourself time to adjust your patterns of speech for the greatest accuracy?

Prevent burnout

◆ How are you preventing your own burnout?

◆ How many quality holidays do you take? Take as many as you can. The faster you run, the more you need recovery time. These spaces are where the real thinking happens.

◆ How do you relax? Remember you're planning for the long haul.

Be sceptical about the latest fads

◆ Are you over-dependent on new management theories?

◆ What is *your* theory of management – does it work?

Break free from the office

◆ Why are you still addicted to office activity?

◆ Does it really improve your performance every day?

◆ How much time would it save if you worked from home at least a day a week?

◆ Do you really need a desk of your own?

◆ Does your own home have enough telephone lines to make tele-working easy? Do you have a home fax with a dedicated line?

◆ Are you familiar with the facilities provided by your mobile phone – and is it fully up to date, with the latest communication features?

◆ Have you considered a satellite mobile phone for remote locations?

◆ What about a completely integrated pocket-sized phone, computer, fax, e-mail, web surfer, diary and word processor? Get one for the same price as a fax machine and run a virtual office from it, use it for conference calls and run the world from a boat, car, train or bicycle.

Enjoy cutting-edge technology

◆ Is the computer you are using sufficiently new and powerful to keep you focused on cutting-edge technology?

◆ Does it have all the latest add-ons for multimedia and communications?

◆ In addition do you have a high specification portable PC – vital for complete virtual working? Buy your own!

◆ Does it have broadband net access and video capability? Indulge yourself and stay ahead.

Be ready for digital disaster

◆ Does your computer have powerful and reliable backup facilities, and do you use them every day?

◆ What about your personal organiser?

◆ Would you be able to survive the theft of all computer equipment in your office and all office backup tapes?

◆ Would your personal work survive a major office fire?

◆ Are you properly protected against viruses in computers, including those from the Internet?

◆ Have you considered using an encryption program for security when sending e-mail?

Keep surfing

◆ Do you have a fast enough Internet connection at home, where you are most likely to learn about new Net technologies?

◆ Do you spend at least one hour a week on the Internet as part of your own professional development?
◆ Have you explored fully new Net areas such as a TV multimedia, and always-on video linking?
◆ How recently?

Wage a war on paper

◆ How fast are you moving towards a paperless office?
◆ Has your paper filing capacity reduced or increased over the last three months?
◆ Have you bought goods from the Internet, and downloaded software?
◆ Have you tried working on shared documents using programs like Net Meeting or made long-distance telephone calls or video links on the Internet?

Keep looking at the camera

◆ Have you tried using a video-conference suite?
◆ Do you know how to really communicate using this medium?
◆ Do you know how to fill the screen with your head and shoulders by adjusting your position and the camera?
◆ Do you look directly at the camera when you are speaking to create eye-to-eye contact – or do you spend most of the time looking at the other person or even at your own image? Your eyes are the window of your soul. Use them. Don't expect results from video links if you seem to be spending your entire time looking away. Appearance is everything in a virtual exchange. Arrive early, practise with a friend, and spend time socialising as you would in a face-to-face meeting. Don't be in such a hurry. Enjoy being together. Create shared spaces.

Be vigilant

◆ How secure is your office from bugging and other invasions of your privacy?

Ten strategies to double personal productivity at zero cost

1. Improve typing speeds/use speech recognition.
2. Good e-mail programme – set up correctly to screen and file automatically.
3. Strict e-mail disciplines, e.g. a ban on most attachments – this protects against viruses and speeds opening mail. And get yourself a programme to block spam junk messages.
4. Phone less, conference calls more, e-mail more.
5. Push videoconferencing and travel a little less.
6. Encourage home-working across time-zones and for writing longer documents.
7. Encourage own diary and travel management.
8. Use mobile dataphones for Internet, e-mail, short text messages and video.
9. Use a bureau to take all phone messages and send to your mobile as text.
10. Encourage out-sourced personal 24-hour technology support.

CHAPTER 2

Urban

Millions drawn to city lights

The second face of the future is urban – the increasing impact and effects of urbanisation, and of other socio-demographic shifts.

POPULATION GROWTH A MAJOR THREAT

Everyone talks about the population explosion, but the graph is actually a straight line, expected gradually to level off. Actually, it is two lines added together: an ageing and shrinking population in many wealthy countries and a juvenile, expanding population in the poorest countries. Population growth will be a major challenge, a fundamental and dominant feature of the future, producing major tensions as poorer countries enlarge at the expense of richer countries with older, dependent populations.

World population, around 6 billion in 2001, will increase to at least 8 billion by 2025, with 95 per cent of the growth being in the poorest nations. These extra 2 billion people will accumulate because birth rates are falling less quickly than improvements in health are made. A huge challenge will be to develop low cost ways to feed, clothe, shelter, power up and water such vast numbers of people without destroying the planet.

Population growth cannot be arrested suddenly without creating all kinds of other crises, with huge population bulges of elderly people in later generations that will dwarf the problems faced in Western countries today. Many, if not the majority of nations worldwide have up to 50 per cent of their population under 15 years old. That is a fact which cannot be undone without the catastrophic death of vast numbers of young adults through plague or world war. Even if not a single baby is born in these nations over the next twenty years, this one age-bulge guarantees a boom in the number of potential parents over the next 20 years.

More than half the world now lives in cities, many vast, primitive and disease-ridden, with poor infrastructure. A large number are megacities – urban areas with more than 10 million people – with vast sprawling areas of crude hovels, slums with poor water provision, no waste collection, poor sanitation, high pollution, traffic jams, power shortages, disease and low investment. Expect this process of urban drift to accelerate. Yet for all this, mega-slums have micro-economies which are often thriving, whether it is a quick-thinking entrepreneur who rigs up electric lines from a lamp-post and starts selling electricity 'by the line' or a boy with a barrow carting food in from outside the city.

Asia has nine megacities: Beijing, Bombay, Calcutta, Jakarta, Osaka, Seoul, Shanghai, Tianjin and Tokyo. Soon four more will be added: Bangkok, Dhaka, Karachi and Manila. Waiting in the wings are Lahore, New Delhi, Bangalore, Madras, Hyderabad, Rangoon and northern China's Shenyang. By 1995 126 million Asians lived in megacities. By 2025 it will be 382 million. By then Asia will contain half the world's population.

Jakarta and the surrounding area has a current population of 12 million, expected to double by 2025. The region experiences frequent flooding, overpumping of groundwater, and air pollution and has high-density slums with little infrastructure. Ten new urban centres in this mass of humanity are planned. In Bangkok most waste water is discharged into storm water drainage systems with little or no treatment.

Welcome to the megacity

The contrasts are bizarre and unsustainable, but getting worse. Take Mumbai (Bombay), a city twice the size of London, its streets clogged with traffic. Every inch of crumbling tarmac is a weaving chaos of cars belching fumes, taxis, buses, lorries and cycles. The streets are so packed with never-ending streams of people that sometimes the only place left to stand is the gutter. The warm, humid air is thick with sulphurous vapours from power stations and heavy industry, mixed with carbon monoxide, carbon dioxide, dust and the heavy smell of humanity.

Along many roads are humble shacks, stacked against each other, packed any old how as far as the eye can see. Cardboard, planks, blue or black plastic sheets, old car tyres, bits of corrugated metal – and rope to hold it all together. The nearest tap is a hundred metres away. Electricity goes as far as the street lamps. A million people every night don't even have a shack to sleep in. They just settle down for the night where their tired bodies collapse on the pavement. They lie as if dead until dawn.

The contrasts are brutal – between the street and the shack on one hand and the rupee millionaire on the other. Flats in Mumbai can fetch more than $1 million with slums almost to the front door. Yet on every street corner you can see CNN in the bars or in the shop windows. However destitute they are, the very poorest always have before them images of a world beyond their wildest hopes, a world of exclusive, unattainable affluence. And it seems unfair. But it also creates a mirage, giving some of the very poorest the hope they need in order to survive.

But we are not just talking about Mumbai: this is the future destiny of cities right across South East Asia. This is life in megacities the world over and it's a challenge. Those not living in megacities live either in other developing urban sprawls or working the fields in subsistence farming, with not much in between.

Gap between rich and poor growing

The gap is getting wider between the richest and the poorest, and many people are falling backwards, particularly in places like central Africa where AIDS, foreign debt and continuous low-grade tribal conflicts/wars are crippling national economies.

In 1965, the income per head of the top 20 per cent of the global population was 30 times that of the poorest. Today the gap is 60-fold. Only a handful of East Asian economies have managed to sustain growth rates fast enough to catch up. Middle income countries – with between 40 and 80 per cent of the income average of all countries – are thinner on the ground than they were.

Gross inequalities are a new form of domination

To have a quarter of the entire human race deprived of basic necessities, such as clean water or adequate food, is shameful in such a wealthy world. Over 60 per cent of the world's population exists on $2 a day or less. Nearly 1 billion are illiterate. Every day around 840 million are hungry or face food insecurity. Almost one in three of those in the least developed countries die before the age of 40. Over a billion people lack access to adequate water supplies. Expect a growing backlash in poorer nations and among activists in wealthy nations. It will be as great as the reaction against every trace of the British Empire seen over the second half of the twentieth century. The debt burden of many of the poorest nations exceeds many times over their entire export value per year and dealing with this problem will be a major challenge to global harmony, despite recent attempts at debt relief. We will look more at this issue later in this chapter.

Foreign aid will be seen as imperialism

Foreign aid can seem a form of imperialism, especially in view of the fact that these same countries are extracting billions more in debt interest than is given in aid. Some countries have economies dominated by development programmes. In places like Burundi a large percentage of vehicles on the road belong to organisations such as UNICEF, Save the Children or ActionAid, a sign of a 'donor-dependent' economy.

Not so long ago I attended a meeting of the World Health Organisation in Geneva Global Programme on AIDS. Every donor nation was represented and they had their own agendas, utterly different from the priorities of the recipient nations. Only six representatives were permitted for most of the rest of the world, including the whole of Africa. They struggled to make their voices heard above the aggressive and opinionated chorus of government ministers from countries like the US, Norway and Britain. It was a poor forum and collapsed, to be replaced by UNAIDS. Foreign aid usually has strings attached and is used openly to control government policies in recipient countries.

It is also extremely difficult for charities to operate in cash-strapped countries without distorting local priorities. A community leader has a shopping list of ten items, ranging from roads to water and clinics. An NGO is offering help with literacy. The help is accepted and a new education facility is built. It may be helpful, but was it the most appropriate next step?

The worst poverty can be abolished in many nations with remarkable speed. Take Malaysia. In 1971 the government developed pro-poor growth policies. At that time 60 per cent were considered below the poverty line. By 1993 this had dropped to 13 per cent, but that was before the 1997 currency collapse. These are brittle and vulnerable economies, but economic growth can do miracles.

---

*Bottom Ten Countries in the UN
Human Development Report:*

(The country at the top of the list is the poorest.)

Burundi
Madagascar
Guinea
Mozambique
Cambodia
Mali
Ethiopia
Burkina Faso
Sierra Leone
Niger

Expect special attention to aid for these countries.

Entering new chaotic markets

A key challenge will be to adapt to megacity culture with its own unique patterns of life and social networks. Expect corporations to penetrate megacities with new products and services and to find new models of management. Image building will be important. Expect a new emphasis on corporate responsibility and global citizenship, with more high profile community action programmes.

Companies operating in the poorest nations will be wise to play down their vast wealth, and to cultivate an image of restraint rather than extravagance, particularly regarding the lifestyles of senior staff. Companies flaunting wealth will invite national criticism, local opposition and labour conflict.

Likewise the challenges for wealthy individuals will be immense, faced by such extremes of affluence and human need. Expect many to turn a blind eye, while others engage in a new wave of intensive humanitarian activity, encouraging sustainable development.

Chinese urbanisation

A major challenge for business will be to push into China at the right time and in the right way, neither wasting huge resources in an embryonic market economy, nor leaving it too late to maintain an early entrant advantage. Hong Kong and Shanghai are doorways.

With 1.2 billion potential consumers, foreign investors pumped in billions between 1990 and 2000, but many companies are finding problems after the initial euphoria. Companies can't afford not to be there but making money will be hard. The theoretical market melts away when you analyse the amount of disposable income. It is still a tiny amount per capita and there is over-capacity in many industry sectors, but now is the time to learn.

One sign of the future is Suzhou industrial park: a new 'supercity' that has sprung up in 70 square kilometres, costing $20 billion from Singapore, creating homes and jobs for 600,000 people. Singapore skills are being used to build water treatment centres, power plants and firstworld houses. Two hundred and twenty Chinese officials have been sent to Singapore to learn how to manage a city 'Singapore style'. We need to learn business 'Chinese style'.

But Suzhou is just a taster of a bigger, sweeping revolution. Yesterday's paddy-field is tomorrow's high-rise. Foreign investment in Suzhou has become unstoppable since the central government signalled a more relaxed attitude in the 1980s. Over 90 projects have already committed $2.5 billion in capital. The community will be clean, ultra-modern and efficient, with luxury lakeside apartments and top-league schools as well as the very latest facilities for industry. Singaporean expertise in everything from fire control to urban planning has been used heavily.

In another industrial zone not far from Suzhou the average income has soared from 1,000 yuan ($183) to 7,000. Du Pont, Siemens, L'Oreal, Acer, Philips and Sharp are building factories fast. Over 200 foreign-funded firms and over 1,000 expatriates have moved into the area. Aggressive state marketing has been helped by the waiving of import taxes, for example to encourage high-tech industry. Expect further revolutionary steps. Three hundred million were lifted out of poverty in 20 years: what of the next 20?

INDUSTRIALISED CITIES NOT DYING

In the 1970s many were predicting the death of cities in Western nations as the middle classes moved out, helped by faster travel and technology, leaving behind the underclass and the black ghettos of many American cities, which would then fall apart through lack of infrastructure. The reality is a mixed picture. Indeed the opposite is true in some places. Take a great city like Chicago: from the air some parts of the city look like a continuous building site as people drift back, looking for quality homes nearer where they work and where the action is.

London alive and kicking

London is also experiencing a magnet effect, firmly placed near the top of the world order in popularity as a place to live. Restaurants and wine bars have multiplied, together with cinema complexes, hotels and nightclubs. London has become one vast work and leisure complex offering the very best of world-class time out for busy executives, round the clock. Those who really want the country had better find a second home, with a London pad for late nights in town. London population has grown a million in a decade.

Expect London's popularity to be threatened as a global centre of financial services, as the network society makes geography increasingly irrelevant. Expect a fierce fight to stay the main player for foreign exchange, and to consolidate position as major European banks switch operations rapidly in and out of countries. Despite this, expect London to keep top position, or near top, in cross-border lending (currently 18 per cent of the global $10,000 billion market). Expect a struggle to retain the world's largest collection of foreign banking offices. The City's net overseas invisible earnings are over £25 billion a year. Expect that to rise in real terms.

Globalised travellers move near airport hubs

Air travel dictates that globalised executives have to be close to a major transport hub, a large international airport. Life is too short to spend six hours travelling either way, to and from an airport you have to use three or four times every month unless you buy your own plane. The 'hub effect' is also true of high-speed train links. Most of the largest hubs for connections are in, or very close to, major cities. Hubs are where profit lies. Control the hub, control the market. Seize landing rights and wipe out the opposition.

Eccentrics drift out but wealthiest keep city base

Once again there is a trend and countertrend. The trend is always for people to drift to cities for the buzz and opportunity of city life: lots of people means money to spend, markets to tap, services to provide. Those who are fed up with city life and are middle class can afford the luxury of being eccentric, going 'back to nature', getting out of cities for a greener life, greater security, lower housing costs, teleworking or pulling whole offices out with them.

However, the wealthiest just live in both. They have two, three, 10 or 20 homes. That is the trend. And there are many more of them too. But you can bet most of these homes will be in different cities. Why else are city residential property prices so high? The fact is that pre-millennialists have to live in or near cities so that they can carry on their favourite and biggest wealth-creating activity: face-to-face meetings. It will be 2020 before the balance becomes dominated by the post-millennial generation who conduct business differently.

And when pre-millennialists get out of airport buildings having travelled halfway round the world, they expect to be at a meeting within a short taxi ride, not to have to take a train up into the hills over a hundred miles away. So this fantasy of life in the green countryside is just that: a middle-class fantasy for middle-income service providers or government officials who are not globalised, and for post-millennialists who can cope with an unending diet of cyber-meetings without getting insecure that their competitors 'in town' are able to sell more powerfully face to face. However, an increasing number of the super-wealthy will have the best of all worlds, with private helicopters and planes that link their offices, homes, hotels and holidays direct, or via big airports and airliners.

Those who are not so wealthy will find themselves increasingly teleworking from virtual offices, but still needing large numbers of face-to-face meetings. While some will drift out of cities, many others will find life demands that they live close to where the real action is, even if many days a month they work at home.

Post-millennialists will fight for the right to work in cyberspace

Post-millennialists will be different: many will succeed, after great effort, in organising themselves, their employers and their clients to run third millennial companies using third millennial technology. Face-to-face meetings out – virtual meetings in.

WATER WARS

One consequence of growing population and urbanisation is water shortage. Elizabeth Dowdewell, Executive Director of the United Nations Environment Programme, has said that the water crisis threatens to dwarf the energy crisis in significance and severity. Water use will be a major feature of foreign policy decisions by 2010, together with carbon trading/global warming.

More people means more pressure on resources, including water. We could see water wars between nations quarrelling over, say, how much water flows into a country's borders down a long river, or how much one country is allowed to pollute another's water supply. Germany and Austria are being sued by the Black Sea states through the EU for polluting the River Danube with more than 100 tons a year of nitrates and large amounts of phosphorus. Algae blooms in the Black Sea have killed millions of fish, and completely wiped out 40 species.

If there is no rain in the Pyrenees, there will be no water in Andorra. If the Caspian Sea dies as a result of one country's pollution, four other nations suffer. Expect many more rows between neighbours up and down river, whether farmers, villages, towns, cities, nations or regions.

Limited fresh water in the world

Only 2.5 per cent of the world's water is fresh and two-thirds of that is frozen in glaciers or on the ice caps. The renewable fresh water on earth (rainfall) is only 0.008 per cent of all global water. Two-thirds of this is lost in evaporation and transpiration. The rest is runoff, available for use. But there is a big mismatch between runoff and population. Asia has 36 per cent of runoff but 60 per cent of the world's people. South America has 26 per cent of the runoff but only 6 per cent of the population. The Amazon river alone carries 15 per cent of the earth's runoff but is accessible to only 0.4 per cent of the world's population. Many other rivers worldwide are too remote to be of economic use, resulting in the non-availability of 19 per cent.

Globally people already use 35 per cent of accessible supplies. An extra 19 per cent is used 'instream' to dilute pollution, sustain fisheries and transport goods. Thus the human race already uses around half the total world supply through rainwater. But water use tripled between 1950 and 1990 as population soared by some 2.7 billion. Population is set to climb by the same by 2030, a 50 per cent increase, but water supply cannot triple again without severe shortages. So something has to change.

To make matters worse, global warming is likely to dump more rain in some places but rob it from others. The World Meteorological Organisation expects 66 per cent of the world's population to suffer severe constraints on water availability by 2025, on current trends.

Groundwater overpumping and aquifer depletion is occurring in many of the world's most important crop-growing areas, including the western US, large parts of India and north China, where water tables are dropping a metre a year. Expect ever longer and more spectacular water pipelines across countries and continents, balancing supply and population.

Many rivers semi-dead in Asia – more to come

Many rivers die for part of the year in Asia as a result of over-irrigation. These include most rivers in India, among them the mighty Ganges, a principal water source for south Asia, and China's Yellow River whose lower reaches ran dry for an average of 70 days in each of the years 1987 to 1997 and for 122 days in 1995.

With urban dwellers set to reach 5 billion by 2025, steps are being taken to switch water from farms to cities. Expect farms in the majority of countries worldwide to be forced to make severe economies in future – thirsty plants instead of thirsty people. In California, irrigation has fallen by 121,000 hectares in the 1980s and a further 162,000 hectares were taken out of the irrigation fields by 2000. In China, water is being switched from food production into Beijing supplies. Three hundred other Chinese cities are now experiencing water shortages.

Expect the 'water factor' to affect every individual in most industrialised cities by 2020, with widespread water metering, 'grey water' systems (for example bath water stored for watering the garden), and a shift in culture to seeing all water as a limited natural resource. It will be as dominant as energy use.

By 2025 the world becomes thirsty for more

By the year 2025, virtually all of the world's economically accessible rivers could be required to meet the needs of agriculture, industry and households, and to maintain lake and river levels. In other words the world will nearly have run out of existing water supplies by the mid twenty-first century. This is a social and geopolitical time-tomb, which will force government action.

The death of the Aral Sea is described by the government of Uzbekistan as one of the most serious ecological catastrophes in the history of mankind, and it took place in a single generation. Originally one of the world's largest lakes, it is now much smaller areas with shores far from the original coastlines, and with hugely risen mineral levels.

Across the world, there has been a serious decline in water quality. South Africa will have run out of rainwater resources in less than 50 years and will have to make fresh water from the ocean. The country is moving from the ownership of local water by landowners to seeing it as a national treasure, while Brazil now sees its vast fresh water resources as one of the country's top priorities for protection.

Coupled with this is coastal pollution. While half the world's population still lacks basic sanitation, 80 per cent of all coastal pollution is from contaminants carried there by fresh water. Water trade is growing. For example Singapore takes water from Malaysia, cleans it and sells it back. Expect many more such inter-nation deals.

Dams will be bigger and more controversial

Controversy will continue to grow over vast dams like the Three Gorges Dam of the Yangtze River begun by China in 1994, which is forming a lake 600 kilometres long, drowning a city of 250,000 people and displacing up to 1.3 million. Dams often force massive resettlements. The World Bank recently reported that 300 new large (i.e. over 115 metres high) dams a year force 4 million people a year to leave their homes, often ancestral lands. Resettlement is usually badly planned and executed.

The Congo River at Inga could supply half the energy needs of the whole of Africa, and only 6 per cent of Africa's potential hydroelectric capacity has so far been harnessed. In theory dams are a wonderful idea: offering free electricity, free irrigation and flood prevention, providing a tourist attraction and leisure centre, fish farming and drought protection. They create jobs, are national status symbols, and prevent global warming.

But dams also change the environment. Constant irrigation can waterlog the ground. Water brings salts to the surface which are left behind as the moisture evaporates, leaving salinated farmland which is increasingly infertile. Fertile silt which used to be carried by floods now clogs up reservoirs. Plant and animal life is lost in a former river, and fish are blocked from going upstream by the dam wall. A study of the Kainji dam on the Niger found that downstream, rice production was reduced by 18 per cent and fish catches by 60 to 70 per cent.

Making more water

So what happens when the water runs short? One answer is to recycle sewage directly back into a large reservoir, after treatment. San Diego is preparing to do just that in a scheme modelled on one 20 years old in Virginia. So is one British water authority in East Anglia. Despite public squeamishness, the re-use of water increased by 30 per cent in one year. The aim is to 'get people comfortable with the idea of drinking treated sewage.' Already treatment technology means that raw sewage can be converted to water 10 times purer than tap water. In the meantime, hundreds of other uses are being found for it, such as irrigation and flushing toilets. Expect sewage re-use to become a big industry. However it is not without risk. Mistakes can kill.

New wells may relieve aquifer problems elsewhere, allowing them to recover to some extent. New dams could increase accessible runoff by around 10 per cent.

Desalination

Desalination is now a proven technology. By December 1995 over 11,000 desalting units had already been ordered or installed worldwide, with a collective capacity of 7.4 billion cubic metres a year. However, desalination today still contributes less than 3 per cent to water use.

Saudi Arabia, the United Arab Emirates and Kuwait accounted for 46 per cent of desalination capacity in 1993, with just 0.4 per cent of the world's population. They can afford to turn oil into water as a gift to the population, who are charged almost nothing, but even Saudi Arabia is now running an economy campaign. It is a lifeline for water-scarce, energy-rich nations, but even if expanded

30-fold desalination could not produce more than 5 per cent of current use. The towing of icebergs, or water-carrying tankers, will also be expensive.

Expect further advances in desalination technology using reverse osmosis (pressurised seawater forced through a membrane, leaving salt behind).

Water conservation – a major national expenditure by 2010

Water conservation costs typically from 5 to 50 cents a cubic metre, less than the development of new resources or desalination. Thames Water in London wastes enough water every day through leakage to fill 500 Olympic-sized swimming pools. Every household could make significant savings – for example on toilet-flushing technology. Only 3 per cent of a British household's consumption was used for cooking or drinking in 2001.

Expect to see a big market in water-friendly devices. Expect the clothing industry to work hard on new easy-clean fabrics, including ones that do not need to be cleaned with either scarce water, or toxic solvents. Expect fabrics and technologies to be developed for air cleaning by 2025 – processes which dislodge molecules of dirt and grease without conventional wetting.

Because farming uses 66 per cent of the world's water, even small savings have a big impact. Expect new drought-resistant crops made by genetic engineering to be seen as water-saving strategies, in effect almost water production plants. Expect new generations of plants that can grow in salt water marshes.

FOREIGN DEBT

Despite huge protest movements, thirty per cent of all African exports were still used simply to repay debt in 2003. At the same time only five nations were providing 0.7 per cent of their GNP for official development assistance (ODA). We are going to hear a lot more about the debt burden on developing countries. A major drive to settle the worst debts by 2000 or soon after was only partly successful. In some countries a huge proportion of all exports is spent paying interest on loans. But what were those loans for?

The fact is that donor countries have for years disguised foreign aid as a boost to their own industry. Most British 'aid' recently has been to provide credits so poor countries can buy British goods. Ninety-six per cent of the debts owed to Britain by the poorest countries are owed to the Export Credits Guarantee Department.

But those goods may not have been top of the priority list in the first place. So, perhaps, a country has an even larger debt because the government decided for political reasons to offload hundreds of Land Rovers built in Britain, paid for out of development money. Even when the vehicles need repair the spares have to be paid for out of scarce currency.

The World Bank lends $24 billion a year, mainly to the poorest nations. But the bank's own review of 83 projects found that half failed to deliver, and many damaged the environment and displaced millions of the poorest and most marginalised people in the world.

The debt problem is so large ($4,000 billion) that when the International Monetary Fund decided to offload some gold in late 1996 to help write off some of the debt (gold gained through interest repayments), it caused a fall in world gold prices.

These debts will lead to added international tension and insta-bility unless dealt with soon. Pressure for change will also come from the millions of men and women in wealthy nations who give sacrificially to relief and development programmes. If the debts were fully cancelled, it would have the same effect as trebling the amount given to relief and development programmes, assuming the released money was used in the same way. Expect more debt-swap programmes, where debt is bought at a big discount from a bank by a development agency and cancelled in return for a commitment by a government to spend national currency on approved programmes.

If the debt problem is not solved quickly, it is not hard to imagine the formation of small groups of economic 'freedom fighters', intent only on targeting banks and other large institutions with terrorist attacks as a protest at what they see as nothing less than economic slavery. It is just part of the bigger picture of growing unease about the impact of wealthy nations and corporations on poor nations and vulnerable populations.

As with other terrorist groups, they will feel passionately that their cause is just and their mission will command sympathy of over a billion people, even if they disagree with terrorist methods.

Even 'wealthy' countries like South Africa are affected

At the Southern Africa Economic Summit I listened recently to an impassioned speech by Nelson Mandela on economic slavery and oppression, 'a greater evil today than 300 years of imperialistic slave trade'. He was bitter about the legacy of debt he inherited on the collapse of apartheid. Countries dependent for decades on the stable prices of raw materials such as minerals, coffee and cocoa beans have also been hard hit by unstable or falling prices, market speculators and 'profit takers'. They know that aid from developed nations is being cut. Poor countries know the greatest wealth will not be generated by aid but by business investment.

Trade rather than aid

In 1996 direct foreign investment in developing countries totalled $244 billion, compared to official inflows of just $42 billion. In 2003 only 1 per cent of direct foreign investment went to the poorest 48 countries. In Africa countries like Tanzania which have changed their attitude to the free market have thrived, while others have wallowed in economic decay. Most governments are chasing each other to embrace low inflation, low budget deficits and the encouragement of private business. But until America and Europe stop blocking their own imports of African goods, and cease dumping their own subsidised farm products, there will be little real progress. Blocking the chance for Africa to export textiles, footwear, leather goods and other labour-intensive products fundamentally undermines the possibility of future prosperity.

Africa was turning around, with an average 4.4 per cent economic growth in 1996 and some exceptional high-fliers such as Uganda growing by as much as 8 per cent. Every one of the countries in the Southern Africa Development Community (SADC) grew in 1996 – Angola, Botswana, Lesotho, Malawi, Mozambique, Namibia,

South Africa, Swaziland, Tanzania, Zambia and Zimbabwe. However these countries are small. Southern Africa is home to only 135 million people and South Africa accounts for 80 per cent of the region's GDP of $150 billion. There are not enough people with cash to spend to make an easy internal market, while the area is isolated from the rest of the world by large distances and poor roads and rail.

However, there is still huge resentment in many senior African leaders, who object to being lectured and cajoled by Western multinationals into lifting currency controls and other measures. They fear the economic rape of their nations by new imperialists who use money instead of guns to invade, take control and rule, to pay little and take wealth out. They see others exploiting the economy rather than themselves, even though many of whom are themselves in highly privileged positions. They particularly resent the fact that they have little choice, it seems, but to allow their economies to become the puppets of globalised power bases. Expect multinationals to find brittle relationships with governments and peoples, which snap at short notice. Currency crises will continue to alarm and frustrate leaders in emerging countries.

RICH CITY LIFE

Life in industrialised nations is also becoming increasingly urbanised. Rich city life creates major problems but of a very different kind. Social collapse, family breakdown, unemployment, addiction, rising crime, the 'black' untaxed economy and the development of an unemployable and disaffected underclass are all major challenges for the future.

Family breakdown

Take the so-called sexual revolution of the 1960s. The dream was of sexual freedom, of free love, but what is the reality in the twenty-first century? Fuelled by increasing mobility, urbanisation and the breakdown of traditional society, we now have a crisis in the US and a situation where Britain spends £1 billion a year placing children in care, the vast majority as a result of family breakdown.

The cost in cash terms of the sexual revolution in the US is $83 billion every year – costs of divorce, sexually transmitted diseases and other factors. Nevertheless something is changing. From 1971 to 1979 the divorce rate in Britain doubled from around 75,000 to 150,000 a year, but since then it has been relatively stable, despite the headlines, with very modest increases each year which have been less and less. In 1996 the divorce rate actually fell. The trend was unsustainable: if the 1970s rates had continued as steeply we would all have landed up divorced before we'd even got married and every child would have been the product of a broken home. Despite all this, couples with children will continue to be a dominant slice of the national make-up, and a key market for a wide range of products and services. Families provide stability and so, in a society which is fragmenting, will tend to win out in the longer term. This is a very important issue for the future.

Marriage less binding than a joint mortgage

'Quickie' divorce and other reforms meant that by the turn of the millennium a joint mortgage between two people was more binding than a marriage ceremony. Divorce hole-in-the-wall machines have been running in Arizona since 1993, with 150 in operation by 1998. Libraries are favourite locations for these instant divorce form dispensers. Once printed they have to be stamped by an attorney's office. The state logs 88 divorces for every 100 marriages.

But the pendulum is swinging in an urbanised, fractured, rapidly changing world where stability is being increasingly prized. For example, the state of Louisiana has introduced a marital covenant that will not permit divorce except under dire circumstances (such as wife-beating, or child abuse, abandonment, two years' separation or adultery). Under the terms of the covenant couples have to declare that they have chosen 'my mate for life wisely' and must also attend pre-marriage counselling.

Community marriage policies

Many towns in America are also adopting 'community marriage policies' in which religious leaders, civic and business leaders agree to encourage couples to get to know each other better before committing themselves. Adopted by more than 60 towns and cities in 26 states, they claim spectacular results, with reductions of up to 40 per cent in divorce rates. Local businesses are asked to sponsor weekend retreats for couples whose marriages are in trouble.

Another sign of the times was a record $1.35 million award by a jury in North Carolina to a woman who lost her husband to his new wife. This was effectively a vast fine, and a huge disincentive to divorce.

Sex everywhere but with a wind of change

Sex is everywhere: the western media continues to break every remaining sexual taboo. That was the trend, but the countertrend is alive and kicking.

The 'True Love Waits' movement in the US has half a million young people who have signed a pledge at public rallies to remain sexually abstinent until marriage 'from this day forth'. The number of women having pre-marital sex is now falling for the first time since records began – 5 per cent down. One follower said recently: 'Sexual liberation is as much about the right to say no. I'm a feminist and I think I'm a lot more liberated in choosing not to have sex.' The movement had taken off in 76 other countries, including South Africa where it has become the 'fastest growing fad among young people', with 120,000 members in just three years, increasing at 5,000 a month.

It's not just the AIDS factor. There's something far deeper here, a profound rethink one generation on from the 1960s dream.

Pro-family – a key government issue

Governments cannot afford to care for societies where conventional families have become rare, regardless of their own philosophies. A sign of the times has been the conversion of left-wing parties in European countries to 'pro-family' rhetoric. The alternative is impossibly expensive. Expect 'family' to become a central issue in the next two decades in all aspects of public policy, with great debate about how to define it. It is a theme that will dominate thinking about reform of the welfare state.

Then there's the Pill factor. Society has yet to come to terms with the Pill. Less than half a century has passed by since sex became separated from pregnancy. It will take another two decades at least for global culture to adjust. The M generation are products of the sixties and seventies because that was the era their parents were brought up in. But very few parents of the 'free love'/'sixties' generation want those same values for their children today. They have lived through it all, done it all, felt the pain, and are hoping for something better for their own children.

M generation will keep the romantic ideal

Most young people are still very romantic. If you ask them what they want out of a relationship most will tell you that their ideal relationship would be with a wonderful person, best friend, companion, great lover. Their ideal relationship carries on for more than a week or a month. Length becomes the test of a great relationship, in an age when people don't expect to stay together unless they are happy. The best relationship in the world goes on and on and on. It never breaks up. And there's increasing curiosity about what makes a great relationship work – just read the agony columns. Expect new specialities to emerge: relationship clinics and training centres, backed with government money.

Expect a new parent training industry as the state starts to panic. Already the British government has announced new plans to provide older mentors for young parents, helping advise on changing nappies, discipline and a host of other things that grandparents helped with before. But these are expensive, partial solutions. A social shift in attitudes will do more, and will begin to deliver by 2020. With it will come some negatives, for example a growing intolerance of single parenting as a positive choice by young mothers. Expect an increasingly vitriolic rearguard action against such things as a return of 'Victorian' stigmatisation.

Many decades to reverse trends fully

Despite the beginnings of a pendulum shift, it will take many decades to reverse the trend for people to live alone. The Department of the Environment in England calculates that 4.4 million new homes will be needed by 2016, 80 per cent for single people.

ADDICTION

Another feature of urbanisation is addiction. Drug dependency is a growing threat to civilised life. Illegal drug traffic is now 8 per cent of the value of all international trade, according to the UN. We tend to think of this as only a problem for wealthy nations but this is far from true. I recently visited people dying of AIDS in Manipur, north east India, on the Burmese border, where 8,000 out of 40,000 people are daily injecting heroin. Every city in Asia now has a significant drugs problem. Each year $1,000 billion is spent on illegal drugs worldwide, most of it laundered through legitimate channels. The figure for America alone is $400 billion. Burma is the largest heroin producer. In South East Asia, Burmese, Afghan and Vietnamese gangs are cultivating 600,000 acres of opium poppies, producing three million tons of opium for heroin a year. Deregulation has made it far more difficult to stop vast flows of dirty money. The Internet will make it almost impossible to track money flows in future. Criminals hate shipping cash, but why bother when electronic pulses can do it all? Laundering is carried out through all the world's main financial centres.

Mexican economy high on drugs

Mexico is so hooked on drugs money that the profits from illegal trade would make the difference for the Mexican economy between boom and bust. The forces of law and order are being subverted, with border guards carrying large quantities in their own vehicles. Mexico's narco-profits are conservatively estimated at $15 billion a year, 5 per cent of GDP. The unfenced 2,000-mile border with the US is so scantily policed that there is virtually free flow. Up to

2,000 lorries a day pass through checkpoints manned by a single customs officer, so inspections are rare.

Mexican gangs have overtaken Colombians as the most powerful drug cartels in America. They supply around 770 tons a year of cocaine, 6.6 tons of heroin and 7,700 tons of cannabis – 70 per cent of all drugs entering the US. Drug cartels are hiring former US military officers at up to $500,000 a year for their expertise on such things as burst transmissions, bug interception and detection and intelligence on highly classified drugs operations.

Drugs in Britain

In 1996 there were more than 43,000 registered users of controlled drugs in the UK, but the real total by 2003 could have been 200,000. Official numbers doubled in ten years. If each drug user spends an average £7,500 a year then the market in hard drugs is worth £1.5 billion to British dealers.

The government says that 20 per cent of all criminals use heroin and heroin users are stealing £1.3 billion a year in property to pay for their habit. In Lancashire that amounts to a loss of £147 for every household. A Department of Health survey of 1,100 addicts found that they had committed more than 70,000 separate crimes in three months before entering treatment.

Expect tough new measures to deal with the problem including 'example' sentences, like the nine months in prison handed out to a woman who gave one ecstasy tablet to a friend at a party. Expect more widespread random testing in prisons, with rewards for those who stay drug free. Expect whole prison wings to become drug-free zones, with extra privileges for those who live in them. Expect new partnerships between prisons and rehabilitation agencies.

Expect more money for residential treatment as surveys show significant numbers of those treated are able to stay off hard drugs. One California study found that every dollar spent on treatment yielded seven dollars in savings – crime losses and prisoner treatment costs for example. The British Department of Health found that treatment reduced shoplifting by heroin users by between 40 and 85 per cent.

A new prohibition movement starts with smoking

Expect a new prohibition movement by 2010, starting with tobacco and attempting even alcohol use in its wake. It will be a countertrend to the growing clamour for legislation of some or all psychoactive drugs. Smoking is rapidly becoming outlawed in the US and we can expect other countries to follow. Smoking kills 419,000 Americans a year and drains $100 billion in health care costs and lost productivity. Second-hand (passive) smoking causes 3,000 lung cancer deaths a year in non-smokers. Workers exposed to second-hand smoke are 34 per cent more likely to get lung cancer. More than 90 per cent of Americans favour the restriction or banning of smoking in public places.

This is no repeat of the temperance movement of the nineteenth century, backed by laws, but a mass movement controlling millions of people through minor inconveniences and regulations. People now have to choose between being a smoker or having a job when it comes to the US public sector, because there is nowhere to smoke at work. Private employers are often just as strict, encouraged by lawsuits blaming passive smoking for ill health.

Britain is going the same way, with smoking banned on buses and underground trains and severe restrictions on main-line trains; a ban in half of Britain's hotel restaurants, and at several football clubs. Some of London's high profile finance houses and banks have bans inside their offices and strongly discourage their employees from smoking outside because of the negative image created. A spokesman for the London Metal Exchange said: 'You don't want a bunch of grubby traders hanging around outside puffing away. It presents entirely the wrong image.'

This creeping prohibition is affecting drug use in the workplace. The majority of large US companies already have drug-testing policies. Insurance companies are insisting on anti-drug measures following the discovery that drug and alcohol abuse costs industry billions a year in accidents and productivity losses. Expect drug testing at work to be a major issue in other countries such as Britain by 2005.

Smoking increases in Britain

The countertrend with smoking has been an unexpected increase in the number of British smokers for the first time in two decades. Since the 1970s 500,000 a year have given up. But in 1997 there were 13 million smokers, a rise of 340,000 in two years. There is a notable increase among high earners, those in their late thirties and early forties. However, smoking in teenage girls 11 to 15 years old is also at its highest level for 15 years, partly encouraged by chic images of models on catwalks holding cigarettes. Expect smoking to fall worldwide in developed countries, with exceptions in particular social groups, but in the poorest nations 'new' wealth will be spent on tobacco.

Increasing lifespans mean people may be careless

One reason that some groups may reject anti-smoking health messages is because of a third millennial fear of living too long. What happens to a generation that watch their parents grow older and frailer without any signs of dying? For many it may even be good news to be able to die fit and bright at 75, having enjoyed life, instead of at 97 with no mind and very little body.

Therefore expect a boomerang effect when it comes to health messages, a kind of double-think – on the one hand an obsession with ageing and with staying healthy forever, and on the other an increasing apathy about personal health. Both will coexist and can already be seen in the same people at different times. Personal health has always been a fairly irrational issue and will continue to be so.

Smoking lawsuits

The pattern is not so much one of new, draconian laws but of a groundswell of public actions. Hence the avalanche of civil litigation which has swamped the American legal system, forcing tobacco manufacturers to set aside $300 billion or more to pay for smoking illnesses over the next 25 years. Seriously large payouts will force a rethink on cigarette promotion far more effectively than bills in Congress seeking to ban adverts. In a world lighting up 15 billion cigarettes a day the compensation packages could be unthinkably large if America becomes a trendsetter for other countries.

It's not just people who want a cut in this mega-deal. Whole states want manufacturers to pay their medical costs for looking after all their people with smoking-related conditions. Florida won $11.3 billion from five companies over 25 years. Such compensation suits come expensive.

The 1997 accord between the industry and 40 states was put in place when tobacco manufacturers agreed to pay $368.5 billion over 25 years in order to get immunity from prosecution in health-related lawsuits. Under the accord, the industry also agreed to accept severe restrictions on marketing and advertising practices, to place much stronger health warnings on cigarette packets and to incur financial penalties if smoking rates in minors did not drop sharply in five years. However the agreement collapsed. Expect the global fight for compensation to continue for at least three decades.

China cracks down on tobacco

China has already banned cigarette advertising and smoking in public places in 71 cities. It has also banned smoking on trains. However, fines are small and the measures are widely ignored at present. Expect that to change. China burns 1.6 trillion cigarettes a year (25 per cent of the population smoke), making it the world's largest producer and consumer. Most smokers are male and smoking kills 500,000 a year, a figure expected to rise to 2 million a year by 2025. On the other hand, the industry has been the largest source of state revenues for 10 years.

Bitter fight coming over dope (cannabis/marijuana)

Expect a bitter fight over dope, which proponents will claim (correctly) is less addictive and dangerous than nicotine, while opponents claim (equally correctly) that the use of other drugs is usually associated with previous use of cannabis. The fact is that the greatest predictor of smoking dope as a teenager is previously smoking tobacco. The fight will be fought one step at a time, with the medicinal use of cannabis top of the list.

New pro-marijuana laws in California have already gone against federal law, which opposes all cannabis use. A number of experiments with the legalisation of cannabis will continue to be watched with interest (and some dismay) by other nations, with a counter-swing against a dope-dominated culture. There is little doubt that legalisation will result in a generation growing up for which cannabis use is as acceptable as smoking tobacco, if not more so. Expect new research to add evidence that cannabis causes subtle but long-term changes in brain activity.

Road accidents and drug use

Expect roadside skin-surface testing for illegal drugs to become common by 2005. Twenty per cent of all people killed in British road accidents carry traces of illegal drugs.

Designer drugs take off

We will see hundreds of new designer drugs in the next decade, acting not only on the brain but also on other parts of the body, for example to prolong sexual prowess and pleasure, to block the symptoms of a hangover or to enhance memory and intelligence. Expect great controversy over memory-enhancing substances and their use in preparing students for exams, and a huge black market in prescription-only sex enhancers, for women and men.

NEW FOOD AND DIETS

Larger and fatter

An example of how health conflicts with pleasure is over food and biomass. Expect to see a battle of the bulge as whole nations get heavier – and heavier. Fat intake is falling in many countries, but not as fast as the loss of exercise. People eating less and less are getting more obese. Expect to see airlines redesign seat widths, even in economy class, as more people fail to squeeze in without risking injury. Car seats will follow.

Slimming boom

The percentage of overweight Americans has risen from 25 per cent in 1981 to over 35 per cent in 2001. Britons, on average, are one kilogram heavier than ten years ago while across western Europe as a whole, 15 to 20 per cent of middle-aged people are obese. For eastern Europe the figure is 40 to 50 per cent among women in some countries. Obesity-related disorders in the US kill 300,000 a year, and cost $100 billion in care. At the same time the US slimming industry has a turnover of $35 billion. Expect this to grow to an international obsession.

Expect a new generation of 'safe' slimming drugs which kill appetite or prevent food absorption, for a market that could be worth at least $5 billion a year in the US alone. Meanwhile the Fatlash movement has grown, promoting the erroneous belief that to be fat is healthy.

Anti-food for fat people

Expect to see a new food industry selling anti-food, or food with absolutely no nutritional value. The first anti-food was a new fat made from molecules which the body cannot digest. This can be used in cakes, ice creams or any other food. It cooks well, but once eaten comes straight out the other end unchanged. Bowel movements become greasy. There are dangers: fat dissolves vitamins, so a diet rich in this anti-food can produce deficiencies, as well as creating a generation of bingeing anorexics who could consume huge plates of food yet waste away to the point of death.

It will be a third millennial irony: one billion people starving or undernourished through poverty and millions of others using scarce resources to make food that they will waste 100 per cent through total excretion. This is the ultimate in gluttony when taken to extremes, yet may be life-saving if it helps arrest obesity in a generation of food addicts who are increasingly immobile, unused to walking more than 100 yards without needing to sit down. Exercise will be an optional extra in tomorrow's world. Why bother to walk when the world can come to your home, or when your home (your car) can travel?

Neurotic eaters

Expect to see growing neurosis, claims and counter-claims, over the safety of food and drink. BSE was the consequence of turning vegetarian cows into cannibals in the name of farming productivity. The memory is still alive of rows over food irradiation. Expect a growth of 'natural' foods and 'natural' packaging as people begin to worry about oestrogen-like chemicals leaching out of plastic, preferring old-fashioned, recyclable glass containers for milk and other products. Expect anti-additive food companies to create entire kitchen environments where nothing 'artificial' ever contaminates what people eat or drink. Expect new ranges of meat substitutes with the right texture and taste for some who insist on it. Expect a rash of new health scares related to vegetarian diets, such as nutritional deficiencies and worries about additives.

Expect more stampedes by consumers from one food to another, following the latest food scare, as happened with BSE and foot and mouth. Expect an end to the worst poultry factory farms, as worries over salmonella poisoning add another chorus to animal welfare campaigns. Expect food retailers and producers who break the rules to be 'punished' by increasingly militant groups, threatening boycotts and intimidation as well as shareholder action. Expect most changes in the food industry to come from consumer behaviour rather than regulations.

Expect confusion over what is safe and unsafe. Millions fled from drinking 'disgusting' tap water only to find that the bottled water they bought at high cost had higher levels of bacteria, was impossible to differentiate from the tap in blind testing studies and, what's more, was contaminated by pollutants from storage in plastic bottles. Expect sales of vitamins to continue their spectacular growth upwards from the current US spend of $3 billion a year, although controversy will rage over what doses should be taken by whom and when – or even if they should be taken at all.

Vegetarians look for new products

Vegetarianism grew 25 per cent, from 4 per cent to 5 per cent of the population, between 1987 and 1996 in the US. Expect the vegetarian market to grow from 5 per cent to 10 per cent of the

US population by 2015, with vegetarians eating some meat products occasionally, and millions of others eating far less meat than today. Red, fatty meat will be particularly unpopular, because of worries about bowel cancer and heart disease. In Britain 40 per cent of people often eat vegetarian foods, ten times the number of strict vegetarians, and the industry is worth £400 million a year.

Expect fast growth for certain 'veggie' products. For example, sales of vegetarian grills and burgers have increased by 139 per cent in five years in the UK, as technology and taste have improved. Expect new meat substitutes such as Arum to seize market share. Arum is an artificial meat-like substance made from wheat gluten and pea protein with the bite, character, flavour and look of meat. It provides a good balance of amino acids from both cereal and pulse proteins.

The latest guidelines, based on findings from the World Health Organisation and other research, suggest that anyone eating as much red meat a day as in a quarter pounder has an increased risk of cancer of the colon and breast. Even those eating an average of 90 grams of red or processed meat a day should, they say, consider cutting back. Expect gene testing by 2005 to be a rich person's way of sorting out whether they need to take the advice or not (see Chapter 5 for more about the implications of biotechnology). The fact is that only a certain proportion of the population need to be careful about animal fat, because only they carry the disease genes. It is the same for many other diet-related conditions.

Poor people eat more meat

While Western societies increasingly turn up their noses at meat, expect the emerging middle classes in many other parts of the world to celebrate their new wealth with increasing meat consumption. The result will be significant net growth in global meat production, at a rate greater than the 1.7 per cent annual population growth. Expect a rise in the proportion of global grain production used to feed animals, from 38 per cent in 2000 to more than 40 per cent beyond 2005. Expect countries like India to embark on another 'green revolution'. India's grain production has hardly increased over the last few years, with grain imports growing for the first time in 20 years. Expect protests in wealthy nations as huge rural areas in places like India are taken over by industrialised large-scale farming, often of genetically modified crops. Concerns will be displacement of very poor labourers and environmental damage.

REPRODUCTION

Children having babies

Every year puberty comes earlier in both boys and girls. This dramatic change is seen most in larger girls, many of whom in some countries are beginning to show pubertal changes as young as nine years old, in some cases at eight. By the age of seven, 27.2 per cent of African-American girls and 6.7 per cent of white girls in America have obvious pre-puberty body changes, for example breast enlargement. Expect to see pre-pubertal changes in even more seven-year-old girls by the second decade in the third millennium. Expect to see personal agonies, such as the one recently over an 11-year-old boy who is father to a child being carried by a 13-year-old girl. Expect to see nine-year-old fathers and nine-year-old pregnant girls. Expect to see more boys of 10 charged with rape, as four were in West London in 1998.

Doctors are having to rewrite the medical textbooks. What is normal, and who needs treatment? Children are being robbed of half a decade of childhood years and are having to cope with huge hormonal and body changes before they are emotionally ready. At present the age of puberty is considered to be a natural and highly personal event, a sacrosanct area that should not be interfered with. But in the third millennium many parents will want to manage puberty in their children rather than risk a full hormonal switch in a child. Society will be forced to take a collective decision about what the 'natural' age of puberty should be, and doctors, together with pharmaceutical companies, will do the rest.

Why is puberty getting earlier?

There are probably two reasons for the lowering of the age of puberty. Firstly it's a simple question of bulk. Female cells produce oestrogen and the larger the female body, the more oestrogen is in the bloodstream, to add to oestrogen from the ovaries. A child who is tall and well built will tend to reach puberty earlier and as we become, with better nutrition, a population of near-giants, so we become a population of child-adults.

Growth patterns are stabilising so this effect should level out. However the second possible factor is environmental. There are a vast number of pollutants in the urban environment with oestrogen-like properties. An example is the string of chemicals which leach out of plastic containers into bottled water or any foodstuff.

One of the biggest sources of environmental oestrogen is food. Soya flour is a well-known source, so much so that a loaf of bread is being marketed which is claimed to reduce or eliminate hot flushes in women around the menopause. But what happens to male consumers?

A sterile generation?

Environmental oestrogens in an increasingly urbanised society may also be the explanation for a catastrophic loss of sperm in men in industrialised nations, where counts have fallen by half in 50 years. In 1992 a study was published in the *British Medical Journal*, summarising the results of 61 studies, going back as far as 1938, involving 15,000 men with no history of infertility. The average sperm count was 113 million per millilitre in 1940 and 66 million in 1990. There is also a significant fall in the motility or healthiness of sperm. On current trends millions of men will be unable to father children because of this effect by the year 2050. If this decline continues at the same rate, sperm counts will be zero in many men within the next 70 to 80 years.

However, these changes mainly affect the wealthiest nations and will not have a noticeable effect on population growth in the next hundred years at current rates. Other factors are more relevant, such as the age at which people start to have children and how many they choose to rear. In any case, men in developed nations will have all the resources of tomorrow's medical technology to help them father children, including cloning (see Chapter 5).

At the same time as sperm counts fall, testicular cancer and prostate cancer are rising. This feminising of the male population is subtle but progressive. Unfortunately the causes are so diffuse, so complex, that it could take several decades to be certain what they are. There are 100,000 widely used industrial chemicals in the environment and 1,000 are added every year.

Age of consent

Age of consent will also be reviewed. In Spain the legal age is 12 compared to 18 in Turkey, 17 in Ireland and 16 in Germany (all these countries have the same age of consent for both heterosexual and homosexual relationships). Expect a great debate over this, and also another where ages for same-sex relationships are different, as in Finland, Greece, Austria and Malta. Expect harmonisation downwards in many countries as law-makers shrink back at the thought of putting older teenagers in prison for having relationships with each other – assuming they object on principle to paying fines.

Child puberty is a double hazard in millennial culture because adulthood is being further and further delayed – settling into a regular job and taking responsibility for others, including children. Apprenticeships and training periods are longer, lasting until the mid-thirties in some cases.

Child-bearing is being put off in almost half of all women in many industrialised countries until after the age of 30. That means a woman may have been ovulating for 20 years before she first attempts to become pregnant.

A strange paradox in the next millennium will be millions of women trying hard to become pregnant in their later years, at a time when millions of pregnancies in younger women are ended in abortion. It will be the era of the precious child, where anything that threatens the health or emotional happiness of a child will be severely frowned on. We see this already in growing public alarm over paedophilia, the safety of children on the roads, the exposure of children to undesirable influences at school and in the media. The new era will be one that worships the little child as a symbol of innocence and perfection in an increasingly tarnished, polluted and self-centred world.

'For the sake of the children' will be rediscovered as a motto to justify more or less anything, from marital fidelity to getting married in the first place or getting divorced, cleaning up the environment, banning cigarette advertising or imposing regulations on the levels of music in nightclubs (for the sake of adolescent children).

Expect parents to become even more preoccupied with child safety, creating cocoons for children in which (they hope) they are totally protected from risk. Fewer children will ride bikes on their own in the park or walk on their own to school. A countertrend will be a new generation of parents who believe children need to be allowed to grow up in the real world, less tied to adults for their every waking moment – parents who welcome risk-taking.

Problems of older mothers

In obstetrics any mother having her first child over the age of 30 is referred to as an 'elderly primip' because the human body is less able to carry a first child at such an old age. The risks are greater in delivery as well as conception, together with added genetic risks. Expect a growing but controversial fashion for women in their fifties to have babies, using donated eggs or their own, held for years in freezers before use. The next two decades will see extraordinary advances in child-making technologies, each of which will push out the boundaries of social acceptability. However, expect to see a reaction against 'playing God' (see Chapter 5), and a growing desire for 'naturalness' in the conception as well as the delivery of babies.

All these biological time-bombs will have major effects in the first two to three decades of the third millennium. And others will emerge. They will fuel a general feeling in growing numbers of people that we are drifting off course and losing control, as a species, of our own destiny.

These issues will feed into much soul-searching over the environment. What kind of world have we made for ourselves? What are these unseen and unknown enemies around us that are making our children grow up too early, grow sick or even die?

THE CULTURE OF THE CITY

Nightclub culture, drugs and deafness

Nightclubs are booming and are a central part of urban life for millions of wealthier teenagers and young adults. Often the music played is completely different from that played at home, highly repetitive and radically different from traditional pop – part of a culture dominated by the use of ecstasy and other similar drugs.

Ecstasy, or MDMA, was first created in 1912 by the Merck pharmaceutical company in Germany and started out as an experimental medicine. Now it has become part of the music. Many tracks are designed to be felt, with ultra-low frequency vibrations forced into the lungs and every body cavity by cavernous speakers which literally move the air. A drug-driven group will have the energy to dance almost without a break from midnight till dawn, or beyond. In part the drug culture is a reaction, an escape from a fast, urban and increasingly chaotic world.

Epidemic of music-related deafness

Expect to see a flood of research papers on deafness. Already the first reports are trickling in of significant hearing damage in some people following a single exposure of several hours at high volume. The Sony Walkman and other similar devices are also capable of destroying the ability to hear certain frequencies, and this can be particularly risky since someone's exposure may be several hours a day without anyone realising. Expect thousands of lawsuits, much like those over tobacco, with people claiming that companies were aware of the potential for hearing damage in the 1980s and 90s but failed to provide proper warnings or regulate the sound output.

Many portable systems can power headphone or earpiece levels far above those from which ear protection is mandatory in factories. It will not be till 2030 that the problem really begins to bite, as the natural ageing process of hearing loss exposes the fact many people began adulthood with subtle but very significant hearing deficits. Expect significant numbers of worst cases to emerge from 2010 on. Expect an effect on music styles and dance culture, with the routine use of wax ear plugs that enable music to be felt rather than heard. Expect the common use of earpieces and earplugs by band members when performing, replacing old, loud onstage monitor speakers as a generation of older musicians find they can hardly hear.

So what will third millennial music be like?

Like previous musical eras, each built on the new technology of the time, third millennial music will harness new tools. We already have the capability to create any waveform we want – but the future will be in reproduction. Whole new ways of creating ambient sound in rooms, so they sound like concert halls or the open air, for example. Three-dimensional sound imaging creates instrumentation that flies through the air and then under your feet, voices that create a sense of total immersion, quite unlike anything ever experienced before – even in real life. These sound constructions will make all existing CDs redundant and every early 2000s album will sound dull, except when played through old equipment used to capture the 'genuine' historic sound. There will also be a music of extremes – every extreme, with many rejecting the results as music at all.

And third millennial films?

Expect to see major changes in the film industry. Expect a flood of remakes with better and more expensive special effects. The films of tomorrow will be dominated by computer-generated virtual sequences where stunt men are replaced by virtuality, and actors become virtual entities. Ever since Eddie Murphy's *The Nutty Professor* and Jim Carrey's *The Mask* actors have been on notice that their services may not be needed in the traditional way. Thousands of feet of film are created with the total computer-generated distortion from reality of the images of original actors.

The three *Jurassic Park* films were popular not just because of good plot and acting, but also because people wanted to see what it would have been like to be around millions of years ago when dinosaurs walked the earth.

Expect, similarly, the creation of hundreds of 'see what it is like' films, creating entirely unique worlds in stunning detail, which will eventually be in trouble-free high-resolution 3D. Large wall-screens in homes by 2010 will mean films that create audience atmosphere or offer vast, even 360-degree, imaging.

And third millennial fashion?

You will see the same trends in fashion, on the catwalks. New designs will be influenced by new materials as revolutionary as stretch lycra became in the 1990s. Fashion parades will also continue to push towards every extreme, with models parading semi-nude, completely veiled, clean, muddy, soaked, icy, body-painted – anything and everything to get attention. But none of this will create popular third millennial fashion. The dominant styles will be different, but by definition mainstream. Expect to see intelligent clothes that change with temperature in colour or texture, and accessories that contain microchips and are 'wired' with functionality, such as belts, hats, glasses, watches, gloves or trainers – for example providing readouts of distance run.

CRIME

Crime and personal security are the number one concerns for the American people, even though crime rates are falling. The fact is that urbanisation creates an environment where crime can flourish. In remote rural villages people still rarely bother even to lock their doors, because crime is rare.

Murder in Latin America

Murders have soared in Latin American megacities. El Salvador is the most violent, with 140 murders per 100,000 people a year. Someone living there for 30 years has an average 1 in 20 chance of being murdered. A man or woman with 100 relatives and friends will go to the funeral of one of them as a result of a murder every seven to eight years.

Murder Rates per 100,000 Population		
	late 70s/ early 80s	late 80s/ early 90s
Colombia	20.5	89.5
Brazil	11.5	19.7
Trinidad and Tobago	2.1	12.6
Peru	2.4	11.5
United States	10.7	10.1

Crime tends to be a juvenile activity, fed by urban decay and social disintegration, that the vast majority grow out of. In America child crime soared by 61 per cent between 1986 and 1995 among 10 to 17 year olds. Arrests for murder in this age group rose 75 per cent. Expect draconian new laws in many countries, with new sanctions against youngsters and their parents as a desperate attempt to knock some responsibility into this age group. Prison populations continue to soar. In Britain the prison service had to try to convert holiday camps and ships into emergency prisons to hold them all. At current rates a new prison the size of Dartmoor will have to be built every six weeks.

Curfews after 9 pm for under ten year olds

Expect the widespread use of curfews in badly affected cities or estates, to cut down on delinquency. More than 100 US cities already require everyone under the age of 17 to be at home by 8 o'clock every night of the week, except on Friday and Saturday when there is an extension to 11 pm. New Orleans police say: 'We've seen a reduction in drive-by shootings involving juveniles.' Teenage deaths from gunfights have halved.

Florida Republican Bill McCollum said recently: 'No population poses a greater threat to public safety than juvenile criminals.' He points out that while US murder rates by adults fell from 1985 to 1995, they rose at the hands of teenagers. Juvenile arrests for all crimes jumped by 67 per cent, and juveniles now commit 20 per cent of all violent crimes. With the numbers of teenagers rising by 15 per cent over the next decade it could get much worse.

Scientific evidence shows overwhelmingly that in industrialised nations, the collapse of the family is one of the most important risk factors in a teenager becoming involved in crime, along with academic failure and personal unhappiness. In the UK most child crime is committed by those from the underclass, those who drop out from school and are often illiterate.

Death of the police force

The first police force was set up by British Home Secretary Sir Robert Peel in 1829, but a revolution is under way. In 1970 the ratio of police to privately employed guards in the United States was 1.4 to 1, but by 2001 there were three times as many private as public policemen. Private security in America costs $90 billion, compared to $40 billion spent on public police. Expect private security spending to continue to increase globally, especially in areas with the greatest contrasts between rich and poor.

In Britain the number of private guards rose from 80,000 in 1971 to over 300,000 in 2003, roughly twice the numbers of police. Australia and Canada have similar ratios. Lawless countries like Russia and South Africa boast ten times as many private as public police.

Russian law and order is chaotic. The interior ministry bill of $1.7 billion a year exceeds by far state spending on health. The new Russian revolution is still young and vulnerable. Expect Russia to take two decades to stabilise, and achieve some 'memory' of stable democracy, sound tax and legal frameworks and an enviable crime record.

SHADOW ECONOMY AND TAX AVOIDANCE

Millions of relatively anonymous people are hard to locate and hard to tax. No one sees what they do. While the Internet provides new ways for the wealthy and educated to evade tax, at the bottom end of the scale the cash economy continues to thrive.

No tax, no regulation. How big is the hole? Prof. Friedrich

Schneider at Austria's Linz University has attempted to measure the shadow economy by examining the amount of unexplained cash sloshing around. He calculates that more than a fifth of GDP in Belgium, Italy and Spain is created in the shadow economy and 10 per cent in Britain, a little more in France and Germany. Compare this, for example, with the Belgian government's own estimate of 3.25 per cent of GDP and Britain's own official figure of only 1.5 per cent. If these figures for Europe are correct then the shadow economy is growing at three times the rate of the official one. Across Europe as a whole the shadow economy probably equals the entire legitimate output of Germany and Spain combined.

Expect far more discussion of shadow economy figures, now that the EU requires them to be included in calculating true GDP and national contributions to EU budgets. Countries are being encouraged to inflate their figures for GDP with estimates of drug running, prostitution and other 'turnover', to boost their case for qualifying to be part of a single currency. In Russia the underground economy is probably greater than the official one. Russian tax inspectors now carry stun grenades, tear gas and an assault rifle. In a year 26 were killed, 74 were wounded and one was kidnapped at work.

Even the British figure of 10 per cent is highly significant. Since most of these cash deals are likely to be at the lower end of the wage scale – say a gardener or painter and decorator – it probably means that two out of five people are involved in the shadow economy in some way as earners, and the vast majority of the population are purchasers.

In Italy and France it now costs a company three times the worker's net pay to take someone into an official job, so there is a huge incentive to employ casual labour on an unofficial basis. In Germany the ratio is 2.3 to 1. The service sector is growing fast in most industrialised countries and so are self-employment and part-time work. As incomes rise, more people are paying others cash to clean, shop, cook and do maintenance work around the house, while the Internet and globalisation are also making income flows harder than ever to monitor. The greater the add-on costs of labour and the greater the regulations, the larger the shadow economy. Thus some measures to increase revenue or protect workers become largely self-defeating.

FEMINISATION OF SOCIETY

We live in a society which is becoming radically feminised. The ultimate in 'female access is everything' is the mixed-sex ward in hospital or the use of male locker rooms by women in the army. According to some, every area should be open in an equal, gender-blind society.

Men are in retreat, labelled as testosterone addicts: dangerous, ill-behaved variants of the human species prone to violence, sexual predatory acts and general loutishness and irresponsibility, the victims of a growing chorus of negative comments and abuse. The patriarchal society is rapidly becoming matriarchal. Female instincts and reactions are set to become the future norms.

If a man is strong he is macho, dangerous, stupid and a typical testosterone product. If he is soft and intuitive, sensitive and caring he is an effeminate, 'boring' wimp. And then some women have begun to decide that romance has died.

A backlash against feminisation has already started. For example, Promise Keepers in the US is a men's movement, associated with the religious right, aiming to encourage responsible male citizenship and to restore men to their traditional role as head of the family. It's growing: 1.2 million gathered together at a series of rallies across the US in 1996 and half a million in a single rally in Washington DC in 1997. It has 400 staff and a budget of $115 million, along with tens of thousands of volunteers. Leader Bill McCartney expressed a crisis of male identity and self-worth: 'Men are more likely to break their marriage vows through adultery, violence and abandonment. Men are impregnating women in record numbers and leaving them to deal with the consequences. It is men, overwhelmingly, who commit most of the nation's violent crimes and overpopulate its prison systems.'

Jobs for the girls

Most new jobs in Britain are going to women – an estimated 80 per cent. Most of the jobs lost are those traditionally done by men. Since the 1970s, female employment has risen by 20 per cent while male jobs have fallen by the same percentage. Tomorrow's jobs require flexibility, teamwork, efficiency – favouring women, according to some feminists. The greatest growth of jobs is in part-time service and leisure industries, while traditional full-time manufacturing workers are a dying breed.

Twenty-five per cent of all men aged 18 to 65 were economically inactive in Britain during the last recession. Some were at college or in training – how else can you survive four major career shifts in a lifetime? Others were unemployed. Others were of retirement age, had taken early retirement or were chronically ill.

The feminisation of society is probably most clearly seen in Japan, a strongly male-dominated culture. From 1975 to 2000 the percentage of women attending a four-year university course rose from one in eight to more than one in four. The female workforce has increased from 32 per cent to over 40 per cent in the same period. Women now have one in 25 managerial positions compared to one in 40 in 1984. Sex discrimination in promotion has been outlawed by courts and 12 female staff of a bank recently won $890,000 in compensation for sexual harassment.

Feminist movements are going to have to think where to go next

Expect men's liberation movements to parallel women's activist groups. Expect gender-role confusion to continue, with a backlash from many women over the negative stereotyping of men. Expect major shifts in corporate culture, especially as populations age, creating skills scarcity. Expect many more women to occupy senior positions. Expect continued agonies over the welfare of children brought up in two-career homes. Expect men to sue women for sexual harassment, intimidation and prejudice in recruitment. Expect men to demand male quotas for jobs and continued debate over whether, for example, male nursery nurses should be allowed to escort toddlers to rest rooms.

Expect the M generation to question some of their parents' ambition when they feel it resulted in damage to them as they grew up. Expect them to organise their own child-rearing years differently, with increasing numbers deciding to sacrifice one or both careers for the sake of longer-term family happiness, although with less income. Men will more often be at home looking after children, and will combine this with part-time teleworking.

Feminisation slow to deliver at home and at work
Despite all the above, feminisation still has a long way to go. Men clean the house, but not much more than they did. There is still a glass ceiling blocking promotion for women in many areas and women still have far less leisure time than men.

AGEING POPULATION

Then there is the ageing population in countries such as Japan, the United States of America and Germany. Expect a significant jump in life expectancy with medical advances in therapeutics and biotech (see pages 110, 213–21).

Babies born in the next ten years will live at least twice as long as those born a hundred years previously. Killers like pneumonia or childbirth are being replaced by rheumatoid arthritis, cancers, dementia, crumbling spines, blindness and deafness. Gone are threats in wealthier nations of polio, tuberculosis and whooping cough. Here instead are fears about dependency, and pressures on the retired to become carers for parents also retired. Grandchildren can find that illness in a parent leaves them caring for two generations above them. Corporations, governments and personal pension funds will need to redo their calculations for future pension provision, with huge consequences for balance sheets.

So are we just adding years to life, instead of life to years?
In Japan the number of children aged 14 and under has dropped to 15.5 per cent of the population from 35 per cent in 1955. By 2010 Japan could have the smallest workforce in the world in percentage terms and the largest proportion of elderly people.

German pensions bill will threaten the EU

By 1997 81 million people were alive in Germany. Of the adults, three were of working age for every one retired. Current predictions are that by 2035 71 million people will be alive in Germany. However the ratio of adults will be only five of working age to three who are retired. The reality will be even tougher. Some of those five will not be working. They will be retired, in training, or chronically sick. Of course it is possible, but unlikely, that the few younger adults left over the next couple of decades will abandon restraint and have very large families. But even if they do, it will not be in time to ameliorate the problem significantly before 2025.

By 2035 there will be 4.5 million pensioners in Germany over the age of 80 – up from just 3 million in 1997. Germany's pension problem could be the undoing of the European superstate. Expect radical measures to deal with it. Britain has vast savings in pension funds for the future. These funds own chunks of land, the finest buildings and parts of national art collections as well as huge corporations. But Germany has little set aside. Germany has been spending not saving. Germany has had one plan only, which has been to get working adults to pay for pensioners. This method lasted a hundred years in a growing labour market with increasing salaries, but no longer.

Who is going to pay the bill?

So who is going to pay? Will there be higher corporate taxes? Sales taxes? Income taxes? Will other European states foot the bill? Who is going to want to live and work in Germany or be eligible for a German pension? Expect relaxation on immigration to provide more workers and drive aging economies.

France will wake up from sleep with riots and demonstrations

France is suffering from a similar nightmare. Its public sector workers are still under the strange delusion that they can all retire on good pensions many years earlier than their colleagues in industry. There will be more rioting on French streets when a government finally has the guts to grasp that nettle.

The French Prime Minister Alan Juppe admitted that 'the figures are terrifying'. A generation ago, five French workers were paying for one retired person's pension. Today there are around 2.2 pension contributors for every recipient, a figure expected to become 1.75 by 2015 and only 1.1 by 2040. By then, if nothing changes, every worker will have to finance 100 per cent of another person's monthly pension payout, as in Germany. France is in a serious situation, having virtually no privately funded pensions. Already 10 per cent of its GDP is spent on pensions and there is no fund of investments for the future. In contrast the UK government spends 5.9 per cent of GDP in total today. The increase in French public worker pensions alone by the year 2015 will be Fr150 billion (in 1993 value francs), equivalent to an extra drain of 1.25 per cent of GDP. By 2015, 25 per cent of France's population will be over 60. Most will be entitled, as things stand, to between 66 per cent and 75 per cent of their finishing gross wages, index-linked, all paid from wages of the working.

Italy also has a severe pensions problem. Already its state pension system is supporting 18 million pensioners with the contributions of only 21 million in the workforce, and is currently running an annual deficit of around $50 billion.

At the end of 1995 EU pension fund assets were $1.5 trillion – a third of US assets, paying for 100 million more people. And what is more, the UK and Netherlands accounted for 75 per cent of that. A staggering 85 per cent of European pensions are 'pay as you go'. If EU pension funds grow 9 per cent a year the assets may grow to $15 trillion by 2020, but the US will have $37.9 trillion at the same growth rate of 9 per cent.

Double taxes for future generations

To sort it out future generations in countries such as Germany will be doubly taxed. Firstly they will have to fork out for the vast numbers of people without a pension who need one today. Secondly they will also be forced to start saving hard so that they don't need a state pension when they get old. In effect that's a double tax. You pay into a private pension for yourself, and into a state pension for those older than you. Unfunded liabilities in pay-as-you-go countries are over $14 trillion. Where will it come from?

One way the money will be found will be by cutting 'safety net' benefits for the sick and the unemployed – a recipe for widespread labour unrest. A French government was recently elected promising to create 700,000 jobs, half in the public sector, cut the working week from 39 to 35 hours with no pay reduction, cut VAT, raise wages and pensions, boost state spending on culture and research – without increasing borrowing, spending more or increasing taxes. Lionel Jospin, the new Socialist Prime Minister, then plunged into a commitment to EMU and busily revised all the above.

So what will individuals do? Firstly some will move out of countries where they are very heavily taxed to pay other people's pensions, and go to live in low-tax countries where they just save for their own pensions. Secondly expect far greater attention to the performance of pension funds and to the levels of contribution.

Moving the Goalposts on Retirement age	
France	both sexes 60, public sector 55 in some cases
Italy	women now 55
Britain	women now 60, rising to 65
America	full benefits age rising to 67
New Zealand	60, rising to 65

Over 65s own most of US wealth

People over 65 account for 50 per cent of US income and 75 per cent of all financial assets. Expect a wide range of new products for the 'grey market', such as super-cruise ships and luxury air trips. Well-off older people tend to be careful spenders who enjoy new experiences and travel. Expect many of the role models used in advertising to age 30 years from the up-and-coming 20 to 30 year olds, to those whose children have long left home. Grey power will be far more visible on the high street, in clothes shops, sports shops, car showrooms, garden centres, travel agents, theatres, cinemas and restaurants.

Personal pension plans and investment funds will be growth markets for those nearing retirement, and high value personal services such as face-to-face banking will be especially aimed at those who are retired. There will be a boom industry in buying life insurance policies and converting them into cash pre-death, aimed at people in a disintegrated society who have no wish to pass on capital to descendants.

Other consequences of the 'grey factor'

Countries short of labour will need to import it from other nations. The increase in immigration from poor to wealthy nations will have a profound effect on the ethnic mix of younger populations in towns and cities, especially when combined with higher birth rates in some existing minority communities.

Expect many older people to carry on working until 75, some enjoying a partial pension at any stage from 45 to 75 so as to take on less well remunerated roles for worthy causes. Expect state benefits to start at higher ages by 2010, 65 years in most cases in countries with a severe pensions deficit. However, those with their own personal pensions will continue to choose to retire, or semi-retire, whenever they like. A key strategy for recruitment beyond 2005 will be tempting out of retirement such people, many of whom may have intended originally to stop work at 55 or 60 at the latest.

Expect an army of the fit and active to become volunteers for hundreds of charities, especially geared to the social and emotional needs of older people whose lives are helped by helping others. Expect these organisations to provide a sense of family, destiny, belonging and personal wealth.

At the other end of the social scale, expect a growing underclass, and a layer just above who work part-time until they drop in their seventies, eighties or beyond, unable to survive on miserable state pensions and out of touch with their children, eking out a pitiful existence doing menial service-industry jobs for almost no pay. They will appear on few statistics, since their earnings will often be informal, in cash and undeclared. No rights, no representation, no benefits, no security. A huge problem will be the growing numbers of people in 20 years' time who failed to invest adequately in pension funds. A separate, traumatic problem will be experienced by those whose funds fail to produce what was promised or expected.

Expect older people with long-term, successful and happy marriages to become a respected source of wisdom on contentment, following a growing recognition that happy, stable marriage is a major predictor of general emotional and physical well-being throughout life. Expect the spread of informal fostering or adoption in wealthier nations of retired people as substitute grandparents by those with children at home, where generations are separated by distance or family tensions. Some of these arrangements will attract state funding because reconstructed family groups save money in care bills.

Expect intensive research into replacement of joints with new prosthetics and joint repair using your own cells. Expect progress in treatment for deterioration of the brain including stem cells to renew brain tissue. Expect new creams and drugs that arrest cell ageing and death. Expect many new health management companies to fight for the lion's share of hospital and nursing home management.

Euthanasia calls will grow stronger

Euthanasia will be a number one hot medical issue for the next three decades. In the first 50 years the population of over-sixties alive will double globally. This issue will not go away. The 'right to die' will be packaged with other issues, including the right of doctors to take a decision to end the life of someone who is unfit to take the decision. Doctors will always have to decide, or a court of law. That's because someone has to decide whether the person is 'of sound mind', has all the facts and is not under undue pressure from others.

Expect to see doctors taking the law into their own hands, high profile court cases and the legalisation of 'mercy death' in some countries. Expect more organisations like the Voluntary Euthanasia Society, Hemlock Society, Compassion in Dying and Death with Dignity to be set up. Also expect to see a backlash in those countries where euthanasia is allowed. Expect to see a compromise where active medical curative treatment is abandoned far more frequently at a far earlier stage, allowing 'nature to take its course' with symptom control measures.

Countries allowing the freedom to kill those who want to die will find the elderly and dependent take this way out as the 'responsible' thing to do. The length of stay in nursing homes will fall, and doctors will become lazy regarding pain relief. People who suffer will be terminated. In the Netherlands hospice medicine is badly developed, as a direct consequence of some of the most liberal euthanasia laws in the world. Deaths from active voluntary euthanasia increased from 1.7 per cent to 2.4 per cent, more than 1 in 50, between 1990 and 1995. However, administering painkillers in large doses (likely to cause death) was the cause of 19 per cent of deaths. One per cent of all deaths were the deliberate killing of a patient by a doctor without the patient requesting it. The criteria for euthanasia now include chronic illness and emotional distress. So now we are in the business of killing people just because they are miserable.

Expect a new emphasis in medical training, not just to cure but to manage death and the dying process. Palliative care is already needed. A recent survey in the US found that most Americans don't have proper access to long-term palliative care and a third die in pain that could be eased. Expect palliative medicine to be a key growth area in emerging nations. Expect new breakthroughs in the relief of pain, and sales of pain-relieving drugs to rocket globally in the next decade.

HEALTHCARE PROVISION

Rationing will produce many moral dilemmas

Health rationing is nothing new but will become far more obvious, with high-profile public debates about whether, say, a 70-year-old man or woman should be given a death sentence because of kidney failure, in order to allow a young child to receive twice-weekly dialysis treatment for the same condition. The number one health issue will be to work out guidelines for 'affordable treatment', having the interests of the whole community in mind. The idea of free or low cost health for all will come under severe pressure. Doctors will be accused of playing God and will blame politicians, who in turn will tell the people (accurately) that health spending has never been so high, increasing far more rapidly than inflation.

I was talking recently to the Health Secretary in the United States of America. She told me that health costs were now increasing so fast that unless something happened soon, the budget of her own department would exceed that of the entire federal government today by early in the twenty-first century.

Expect growing concerns under the Bush administration about whether the Medicare budget is sustainable without bankruptcy by 2010. The Medicare budget leapt from around $50 billion in 1985 to well over $150 billion in 1995, with estimates in excess of $400 billion by 2005. Fraud and mismanagement is costing up to $23 billion a year in a complex system settling 80 million claims a year.

Expect a global shortage of trained doctors and nurses by 2005 with wealthy nations having to plug labour shortages by tempting care professionals trained by poorer nations to emigrate. This 'theft' of care teams from the countries needing them most will cause huge hardship in emerging nations. Meanwhile trained professionals in wealthy nations will continue to leave their posts for other careers altogether, blaming poor pay, overwork and low morale.

Health will therefore be a dominant industry in wealthy nations, and increasingly so in aspiring nations, with significant global growth throughout the next century. Doctors, nurses and other health-care professionals will continue to be needed in increasing numbers, despite robotics and other technologies. People power will continue to be the main source of care delivery, whether it's helping someone turn in bed, or listening to someone in distress who is feeling unwell and afraid.

Emotional ill-health will be a growing issue from now to 2020, with depression and anxiety increasingly dominating many people's lives to the point where normal coping mechanisms fail. The history of Western civilisation has been that as people grow more affluent, and as their lives become healthier and easier physically, their stress levels rise and they become more and more emotionally fragile. Hence the US boom in therapy and a host of other services designed to care for the emotions. Expect 20 per cent of all men and women alive today to need formal psychiatric help at some time in their lives, up from one in six women and one in nine men in the UK today.

Expect a new medical specialism to appear, linked to human happiness. Happiness will become a focused theme of the early third millennium. Since improvements in material wealth seem to produce no discernible increase in national or personal contentment or fulfilment, where is happiness to be found? New designer drugs will be increasingly used by doctors to elevate mood or control anxiety, as a partial solution to a growing crisis. Widespread pre-scribing of these mind-altering substances will be a back door to legalised drug intoxication.

New epidemics

One consequence of the increasing population is a growing risk of global epidemics. We are already seeing rapidly changing viruses emerge in different parts of the world. Every time a new person is infected there is a small risk of a significant mutation. As the world population increases, so the risk of mutation increases. High mobility also encourages spread. We have no medical protection against viral plague, no equivalent of penicillin for viruses.

New bacterial plagues are also being seen, for example the flesh-rotting Buruli outbreak in Africa which infected 6,000 in the Ivory Coast alone in 1996, stripping away skin to reveal gangrenous flesh. The only treatment is surgical excision. Expect to see a number of new viral threats by 2020, and intensive research into anti-virals. Expect major breakthroughs in the use of gene technology against flu, the common cold and other viruses by 2010.

A new global threat

AIDS is just an example of these new viral threats, and is now out of control in many of the poorest nations. We are seeing one new infection every 15 seconds, with around one in 200 adults alive at the end of 2003 carrying the virus. Ninety per cent of all new infection is among heterosexuals. In 1996 more people died of AIDS in the US than in the entire ten-year Vietnam war. Some African countries are reporting that one in five of all adults carry the virus.

But the greatest impact will be in the Far East. India now has more cases than any other nation. 2.5 per cent of Mumbai adults are carriers, with 1000 new infections a night. By 2010 India could have more HIV cases than the entire world today. Neighbouring countries will follow, and then China. Expect wealthy nations to invest more in prevention as part of developmental aid. Prevention works and it is the only answer.

HIV treatments for rich not poor

HIV protease inhibitors show promise, but are toxic and expensive. Latest AIDS therapies can cost more than $12,000 per person per year – science fiction for a country like Uganda with only $3 a year per person to spend on health. Expect a 'morning after' pill for HIV in less than five years, with worries about partial efficacy and side effects. Expect no effective vaccine for at least another decade. Once such a vaccine has been developed, expect huge pressure on the patent owner to make it available globally at cost. For this reason, drug companies will continue to pour more money into HIV drugs for wealthy nations, than into vaccines. Expect huge consequences following the decision in 2001 by drug companies to accept legal defeat in the fight to stop poor nations making copies of AIDS medicines. It will be part of a global attack on intellectual property rights, where those rights seem to enable rich companies to exploit poor nations.

SARS: Threat of new viral plagues

In 1918–1919 Spanish flu swept the world killing 30 million people. For years scientists have warned that another virulent strain could mutate into reality at any time. In March 2003 a strange new virus swept across parts of China and Hong Kong, killing 14 in every hundred from a severe atypical respiratory syndrome (SARS) – sudden pneumonia. With no treatment or vaccine the virus was rapidly carried around the world, causing alarm and huge efforts to contain spread. SARS was a wake-up call: expect great effort to develop better antivirals as a matter of global survival.

CHALLENGES TO MANAGEMENT

Megacities
- Do you need a strategy for seizing market share or developing new markets in megacities?
- Who understands megacity culture in your organisation?
- Who is advising you on big questions, such as to what route to use or what speed to go at in regard to emerging countries such as China or India?

Maintaining appropriate differentials
- Expect pay scales to come under increasing scrutiny, with growing questions over vast differentials, not just in the same country and the same company, but between nations in the same company. Do you have a global pay policy which can be justified?

Relocation policy
- A key challenge for many corporations in the next decade will be whether to relocate – and if so, where? Expect a flow in and out of the central areas of big cities of companies with household names.
- Are your offices the right size and in the right place for rapidly changing circumstances – including the need for easy access to a global travel hub?

Water restriction
- What effect will water restrictions have on your business?
- Do you have added value here – e.g. making distilled water while smelting aluminium?

Family and relationship issues
- Have you considered how your company can help those who are married to stay happily married?
- Have you reviewed areas of policy for their 'family value'?
- Have you factored in the loss of productivity and increased staff turnover from family break-up?

◆ Have you thought of creating family-friendly policies as a means of attracting and retaining staff?

◆ Is your company providing an appropriate level of child care or flexible enough options to retain key female staff?

◆ Is your company providing fast career paths for high-flying women?

◆ Do your female staff agree or is there a glass ceiling?

◆ Have you defined and dealt with sexual harassment?

◆ Is there a written and understood policy?

Unemployment and insecurity

◆ What kind of staff training programme do you have?

◆ Do staff feel that even though their jobs may be insecure, they are becoming steadily more employable elsewhere as a result of working for you?

◆ What retraining and consultancy/advice packages do you offer those who become redundant?

Addiction

◆ What is your company policy on addiction and how is it defined?

◆ What is the policy on intoxication at work and how is it measured?

◆ Are employees screened for mind-altering substances, including alcohol, if they are in sensitive situations where the health and safety of others may be at risk?

◆ What sanctions are applied to those who test positive or are intoxicated, and do all employees understand what the sanctions are?

◆ What is the company smoking policy and is it being applied consistently?

Diet and weight

◆ What is the company policy on nutrition – e.g. canteen/restaurant – and has it been reviewed recently?

Crime

◆ How would your company survive the theft of every piece of computer equipment at a key site, including all tape streamers and other backup storage devices and related media?

- What protection do you have against corporate spying, electronic bugging and other information losses?
- How secure is your business from attack?

Ageing population
- Have you reviewed retirement policy in line with demographic change?
- Have you considered abolishing a compulsory retirement age?
- Is your recruitment ageist?
- Have you considered changes in job design and hours to attract older people who may stay longer and be more loyal, as well as being more mature and experienced?
- Is your pension provision adequate, and if not, are your employees aware they may need to top up with voluntary contributions?
- Are you ready to move out of countries where social costs are going to soar, or to enter growing 'grey' markets?

Health provision
- Is your company's healthcare provision in line with what others are about to provide, in the light of the reducing role of free state health care?
- Does your company offer support for staff with emotional problems, as a means of enhancing productivity and loyalty?
- Has your company taken health promotion seriously, particularly regarding AIDS in high risk countries?

PERSONAL CHALLENGES

Transport hub – how close are you?
- How important will air travel be to you in the future, and are you near enough to a major international airport? Life's too short for frequent fliers to spend several hours driving to catch a flight, when it takes less time to travel across an entire continent.

Happy home life – competitive edge

◆ If you are married or in a long-term relationship, how important is its success to you and are you investing enough in it? A happy home life gives you a strong competitive edge. Domestic disaster will wipe you out emotionally and physically.

◆ Where did you last program in some quality time for your most important personal relationships?

◆ What about your oldest and most faithful friends?

Family issues

◆ In view of the rising infertility problems in many countries, how long dare you leave it before beginning to try to start a family?

◆ Where does all that fit into the rest of your plans for the future?

◆ Are you talking to your own children enough about the pressures they feel to take drugs or to experiment with early sexual relationships?

◆ Are they getting enough quality time with you?

◆ How are they coping with the fast, urbanised life you lead?

Tobacco, alcohol and other drugs

◆ Is your use of tobacco, alcohol or other drugs holding you back?

◆ Are you happy about that?

◆ If so, what are you doing about it? There are many other ways of winding down and relaxing.

◆ What would happen if you were randomly tested for drugs or alcohol at work?

◆ Is there a risk it could affect your future?

◆ Are you able to smoke freely when you want and where you want at work?

◆ If not, how does it affect your performance when you want a cigarette or when you have to keep leaving the office to go and have a smoke?

Feminisation of society

◆ How comfortable are you with the feminisation of society?

◆ Do you feel secure in your own role as a man or a woman in the workplace?

- How would you like attitudes to change?
- What steps can you take now in your own situation to encourage this?

Planning to say goodbye to work

- When would you like to retire?
- How will you use the time?
- Is your own pension plan adequate for the future?
- Is there enough of a contingency there to cover periods of unemployment, when contributions will be hard to make, or unexpected early retirement?

Elderly relatives – what's the plan?

- How would serious illness in an elderly parent affect your plans?
- What do you hope will happen when you become old and frail yourself?
- Is that what you are modelling to your children with the generation above?

And finally . . .

- Do you feel comfortable with the gross inequality of wealth between the richest and poorest – and if not, what are you doing about it?
- Have you installed any water-saving devices in your own home?

CHAPTER 3

Tribal

Identity crisis: conflict of culture and conscience

Tribalism is the most powerful force in the world: more powerful than any national army. Tribalism is what makes you and me different. We tend to associate tribalism with African tribal wars in places like Malawi, Rwanda or, more recently, Kenya, but tribalism is seen in any group of people who agree to belong together and tribes can be both positive and negative. The greater the globalisation, the greater the tribalism, particularly when one tribe is so dominant – America.

That is not to say that the end of the sovereign state is at hand. Tribalism simply means more sovereign states, smaller units of jurisdiction, while universalism dictates that larger issues such as defence will tend to be handled regionally.

The *Economist* magazine wrote recently: 'The virus of tribalism risks becoming the AIDS of international politics – lying dormant for years and then flowing up to destroy countries.' We see tribal conflicts everywhere, whether in the Middle East (future flashpoint), Bosnia, Kosovo, Northern Ireland, in Spain with Basque separatists, in Canada with Quebec separatists, in Chechnya, Rwanda, Kashmir, Indonesia or between black and white in America.

Madeline Albright, the United States Secretary of State, reminded Europeans recently: 'From Bosnia to Chechnya, more Europeans died violently in the last five years than in the previous 45. From Serbia to Belarus.' Wars of today are mainly between family and family, and set street against street, town against town. These are not wars between nations but wars inside nations, or wars fought in new ways by an informal enemy whose greatest weapon is fear.

TRIBALISM WILL DESTROY EUROPE

Tribalism will be the downfall of Europe. A United States of Europe has no chance of being a reality for the next hundred years without some kind of imposed authority. In Europe we have nations who can't even hold their own people together. How can they all merge as one? We don't even have a President.

Another major problem is in defining what Europe actually is. It was all very clear before the Iron Curtain collapsed, but now? The old cluster of established industrialised nations is being broadened rapidly, so that the community is an ever-larger collection of very different economies. How can all these be welded into one? The old-style European Union is dead.

Language identity

We see tribalism in the resurgence of language identity. For example, Gaelic in Scotland was more or less a dead language 15 years ago. Now it is spoken in shops, radio programmes are in Gaelic, it is taught in schools and road signs are all in both Gaelic and English.

Then there is French radio. There is a strict limit on the amount of English language music that can be transmitted on air. The medieval French tongue Langue d'oc is experiencing a revival. The first Langue d'oc school opened in 1979; now over 1,400 pupils receive tuition in the language, backed by state funding.

Language is very important: it helps express our national and cultural identity. Language preserves ancient literature and poetry as well as songs. Language communicates who we are – even the accent in which we speak our mother tongue reveals our tribe. This new love of old languages is a direct reaction to the almost overwhelming threat of English, which is used in 60 per cent of world broadcasts, 70 per cent of world mail, 85 per cent of international calls and 80 per cent of all computer data.

In Russia there is a ruling in some cities that business signs over shops must be in Russian as well as English. Belgium is a country split between Flemish and French. Switzerland runs on four languages: French, German, Italian and Romansh.

There are 329 languages spoken in the US. Out of the total population of over 260 million, 198 million speak English at home, 17 million speak Spanish and the rest speak one or more of the other 300-plus. On 1 August 1997 Newt Gingrich, Republican Speaker of the House of Representatives, said: 'This is a level of confusion which, if it was allowed to develop for another 20 to 30 years, would literally lead, I think, to the decay of core parts of our civilisation.

Despite all this, the overwhelming evidence is that ethnic groups have children who adopt the dominant language of the country they live in. Parents have trouble persuading them to keep speaking the domestic mother tongue.

Language protects culture

Mexican Nobel Prize winner Octavio Paz once wrote: 'With every language that dies, an image of mankind is wiped out.' At present between 6,000 and 7,000 languages are spoken. A third of all languages will disappear within the next 100 years. Only 600 languages are considered secure. Expect more vigorous efforts to preserve those that remain. Success will depend on parents encouraging languages at home and school.

TRIBALISM FEEDS TERRORISM

Terrorism thrives on tribalism. Expect to see terrorist groups multiply in the third millennium, the products of tribalism and single-issue activism. Most terrorist groups will continue to be small, fragmented, mobile and short-lived. They will tend mainly to be local rather than globalised with informal networks, using new technologies to frighten, sabotage and attack for the sake of a cause, seeing themselves as moral freedom fighters.

Expect to see a shift from suicide car bombings, random shootings and remote control explosions using the tools of the mid to late twentieth century. Tomorrow's terrorists will be interested in things like germ warfare agents that can threaten city or countryside yet be carried in a briefcase. Recent Sarin nerve agent attacks on the Japanese underground railways are the style of things to come as will be media-grabbing outrages. Terrorists will seek greater power for less effort, using more automatic weapons, and heat-seeking and other kinds of stolen missiles, readily available since the collapse of communism.

The collapse of the World Trade Center in 2001 set a benchmark against which terror groups will measure success. Never have five men with box-cutters demonstrated such power. Angry groups of many kinds will devote great efforts to similar acts – perhaps flying into a nuclear power station, or contaminating a city with a 'dirty' bomb, explosives mixed with radioactive waste. Suicide attacks will be seen as noble and courageous.

Terrorism becomes a way of life

Expect to see economic and anti-corporate terrorism directed at things not people: the spiking of more food products in shops, the cutting of optic cables, damage to huge satellite dishes, the creation of computer viruses and spread of animal viruses like foot and mouth. Expect more tribe- and state-funded terrorism as conventional wars become more difficult to fight.

Expect states to spend more on anti-subversive activity. Security forces will use ever more sophisticated spying technology to track down and destroy terrorist groups, and in so doing will violate the privacy of many innocent citizens. The same forces will also target drugs activity and globalised crime syndicates. Terrorism will become an accepted part of third millennial life, as an irritant rather than a major feature.

TV makes big wars harder to fight

Live TV pictures and videophones will make wars harder to sustain because their horror will be seen so close to home. The Iraq war, Afghanistan, Bosnia, Kosovo – conflicts have become more difficult to manage from the public relations point of view. The media brings home to us the absurd and obscene contrasts in our world. It focuses on violence and discord, rewarding protest groups and tribal factions with huge coverage, hijacking national agendas, and disturbing the sense of well-being of millions.

THE DEATH OF NUCLEAR WEAPONS

Conventional weapons such as land mines have been in great demand while nuclear weapons have been dying. During the Cold War tens of thousands of nuclear missiles were trained by Russia and America on each other's largest cities. By 2001 START-1 had reduced this to around 6,000 long-range weapons each. The new rapport between Presidents Putin and Bush will result in further rapid reductions in nuclear warheads – with continued worries about safe disposal. Expect hundreds of redundant cruise missile launchers to be refitted as satellite launchers for a global network of low-flying telecom satellites, creating total radio access across the earth, with ultra-fast data transmission for mobile networking.

However, the surplus missiles are a great temptation to the crime syndicates which control much of Russia's GDP. Russia will be a major source of surplus arms sales for the next twenty years, selling to virtually anyone with money (with the exception of nuclear sales or attempted sales which will be controlled by the Mafia). Expect to see some major nuclear scares over the first three decades of the third millennium as countries or groups claim to have got hold of nuclear weapons or material, or to have developed their own, and threaten to use them. Bluff and counter-bluff will threaten regional security. There could be some risky stand-offs.

Take a country like India, threatened by border problems with Pakistan and disputes over Kashmir. It is not difficult to imagine a new situation in the next few years where one side creates a crisis so serious that the other feels the only appropriate response is to threaten the use of a small tactical nuclear warhead. Expect an accelerating nuclear arms race involving India, Pakistan, China, Iran, North Korea and others, in a bid to protect national security.

No nuclear warhead has been used in war for 50 years. Expect someone to threaten it somewhere and massive international confusion about how to respond. Do other countries threaten to go to war against a nuclear weapon-using nation, if a warhead is used by such a country in self-defence, after repeated warnings to an aggressor? How would such a war be waged? How do you counterstrike against an invisible terrorist group? What happens if a nation or group threatens again, and explodes another warhead? Do the collective opposition then decide to let off a warhead themselves? Countries may have only days or hours to debate these issues when the situation emerges.

Many treaties but do we believe them?

The Nuclear Non-Proliferation Treaty has been given new powers to hunt out secret bomb-builders. Compliance is always a test of trust. As the nuclear tests carried out by India and Pakistan in 1998 showed, nuclear capability can be difficult to monitor. This applies even more to biological weapons. Signers of the new Chemical Weapons Convention (88 countries have ratified the treaty out of 165 signers) agreed to allow instant inspections as well as routine checks. Even so, hiding chemical or biological facilities is not difficult. New nuclear technologies able to detect even minute amounts of radioactivity in soil, air and water should give greater security about clandestine nuclear bomb-making – in theory.

Russia and China are still spending huge amounts to protect themselves from invasion – or be ready to attack. Together they spent $84 billion in 1995, on combined armies of 4.5 million equipped with 25,000 tanks and 8,000 combat aircraft. A new agreement on their common borders should reduce the huge build-up on either side as part of steps to working more closely together.

WORLD MILITARY SPENDING WILL FALL – AND RISE

World military spending fell in 1996 to $811 billion – 60 per cent of the 1987 peak – to the lowest since 1966, with a fall in jobs from 17.5 million to 11.1 million. American spending halved in a decade. There'll be lots of news about ultra-smart weapons and near-empty battlefields fought over by unmanned drones, cruise missiles and other technology. But the fact is that most wars today are guerrilla wars fought wall by wall and house by house, ethnic conflicts or terrorist attacks; mucky wars where tanks park themselves inside the compound of a large children's hospital, where civilians are caught up in bombing attacks.

There is no such thing as a designer battlefield, except the areas where test exercises are done over and over again. Real battles are fought in shopping precincts, around public libraries, by ancient stone bridges and in fields of corn. The JSTARS positioning system can now pinpoint every vehicle in fog or darkness in a 200 square kilometre area. Fine in a rolling landscape of fields or desert, but smart wars get bogged down in real urban battle.

National arms factories will continue to overproduce

Production and company ownership is still nation-based. There are no multinational Unilevers, IBMs or GlaxoWellcomes in the defence industry. National champions are protected by governments that ban foreign takeovers, protect against imports and maintain capacity, so that nations can survive war. For example, Britain found that Belgium would not sell them shells in the Gulf War. But modern weaponry is incredibly expensive to develop. A Lockheed Martin F-16 fighter plane costs $30 million today, while the F-22 will cost more than $100 million each. America is now the only western country able to afford internal development.

Domestic consumption is too small to justify investment in peacetime. That means weapons have to be sold in vast numbers, by producing nations such as America, Britain and France, to whoever wants them. But buyers are often using these weapons for internal repression of a terrible kind, or to invade neighbours. This is hardly a clean industry. One solution is for nations such as those in Europe to combine forces. Expect more partnerships within NATO, although national overproduction will continue.

The British, US and French economies will suffer as the arms trade shrinks. The UK alone employs 415,000 people in making weapons for war. It is described politely as a defence industry. This is an industry which produces offensive weapons, designed to shoot down, destroy or otherwise incapacitate an enemy. If an enemy is seen as the aggressor, then action is seen as defence. But many of the weapons sold to other countries are used aggressively.

Expect increasing unease in developed countries about the building of national economies on the sale of 'death machines', and growing ethical concerns among the workforce as well as shareholders. The argument until now has been that if we don't sell, someone else will. This argument will not be so persuasive in the future.

Landmines and other messy weapons

So what happens to all this production? Either the world fills up with more and more weapons, millions of rifles, tens of thousands of missiles, or for every weapon made another has to be decommissioned. In practice two things happen. Firstly, weapons move down the arms chain towards the bottom of the pile, into the hands of the poorest (and often most unstable) nations where they are often used for internal repression rather than national security. Then arms fall into the hands of gangs and criminal groups, the lowest layer of all. This happened on a grand scale in Albania during the spring 1997 riots triggered by the collapse of pyramid selling schemes. Over a million weapons were stolen from government forces and the police. It is also a major problem across Africa.

Secondly, weapons are lost or unaccounted for. Landmines are a prime example. Tens of thousands of square miles are uninhabitable because of the indiscriminate use of anti-personnel devices which will remain dangerous for at least two decades. Over 110 million mines have been scattered and lost, affecting at least 70 countries. Cambodia, Angola and Mozambique are among a number of countries which have been severely affected, with large numbers of civilians killed or injured every year, including children. A third of all the land that could be farmed in Afghanistan and Cambodia has been rendered unusable. Worldwide, a million people have been injured or killed in 25 years. There are a further 100 million land mines in military stores.

After much campaigning, the British Labour government announced that it would phase out all stocks by 2005. Many other countries declared support, with the exception of America (worried about verification and compliance), Russia and China. Nations involved in border disputes rely heavily on no-go areas created by extensive mining, and use small anti-personnel devices to prevent mine detectors getting anywhere near larger anti-tank devices. Expect many countries such as the US to continue to use anti-personnel devices even when they have agreed a ban, by adding highly sensitive anti-people booby traps to larger anti-tank mines.

Expect to see today's high-tech weaponry used by 2010 in many developing countries against neighbours and internal threats.

TRIBALISM BROKE UP THE SOVIET UNION

Tribalism, expressed as nationalism, broke up the Soviet Union and will threaten Russia as it is known today. Tribalism will affect the future of countries like the Ukraine, suffering with the aftershock of dealing with inflation of 4,735 per cent in 1993 and an economy that shrank 10 per cent a year. Anti-Russian feeling will continue to find expression in the era of President Putin.

The 12 countries of the old Soviet Union are drifting further and further apart and the new Commonwealth of Independent States (CIS) has failed, whether as a military union, a currency union or an economic union. So long as Russia continues to try to dominate, the CIS members will continue to be forced either into further economic interdependence on the old mother state or, as many of them are, into looking west, possibly as far as trying to join NATO – an act the Kremlin once described to be as dangerous as 'playing Russian roulette'.

The Soviet economic order has gone for good. In 1990 you had to queue to buy bread. Now people drive around with mobile phones. Most countries enjoy open, democratic government. Most state-owned companies have been sold or liquidated. Privately owned business now accounts for more than half of the region's GDP. Prices have been freed and currency restrictions and trade tariffs lifted. A return to communism would itself trigger a people's revolt. Albania aside, this has all been achieved with remarkably little unrest, despite inflation rates which peaked at up to 1,000 per cent before stabilising at around 30 per cent.

Foreign investment is picking up, especially in countries like Poland and Hungary. The European Bank for Reconstruction and Development estimates that $45 billion has been invested in the region since 1991 and the region's stock markets are becoming popular. Expect westernisation to accelerate, and an increasingly wealthy middle class to begin to travel across Europe frequently.

The future of the former Eastern bloc

Much of the region will remain in deep recession for some while. In half of the economies in central and eastern Europe output has fallen by 50 per cent since 1990 – catastrophic, since the original levels were already low compared to the rest of Europe. The entrepreneurial spirit in many communities has been crushed.

It will take another decade for these Eastern bloc countries to settle down economically, and even then there will be a huge gap with the rest of Europe. It will take two decades for democratic traditions to take root. In many of these countries there is a political immaturity which could lead to unrest. Some democracies are only skin-deep, brittle and fragile. Unless there is a great leap forward in economic growth, growing dissatisfaction could lead to riots, civil disobedience, internal military action or worse. Expect wealthy nations to take positive action to reduce the risk of this, including early inclusion in a broader, redefined European Community. New tax systems, regulatory authorities, laws and other components of civic life need to be introduced and adapted, and they will be.

All this will add to a great disturbance of the original simple concept of a western European Economic Community if most or all of these central and eastern countries become part of the whole. But enlarging the EU from 15 to 25 countries will change it forever. For example, unanimous decisions will become impossible. The combined GDP of 11 new applicants is barely 4 per cent of the current EU total, yet they would form a big chunk of the votes.

Seventy-five million new EU citizens will create a community of almost 500 million people, a $9.5 trillion economy. But 45% of EU budget will be subsidies of new entrants. Poland will take 40 years on current trends to reach UK wealth and huge numbers in the 10 new nations earn less than $450 a month.

THE EU – SUM TOTAL OF EVERY COUNTRY'S NEUROSES

The idea of a united European order is not new. It was the foundation of the Holy Roman Empire which lasted from 800 to 1806, extending over large areas of central Europe including what is now France, parts of Italy, Germany and Belgium. The trouble is that when standards on such things as health and safety are combined there is a tendency to push for the highest rather than the lowest.

The result is a sum total of every country's neuroses – on how cattle should be slaughtered, or how clean water should be for drinking, or whether double-decker buses are safe. Every country has its own petty neuroses. These things become part of national culture, dominating chat shows and parliamentary discussions. The British are deeply shocked by animal cruelty, so they want to ban the transport of live animals and fox-hunting – as well, presumably, as bull-fighting in Spain. The Spanish have been neurotic about the safety of British beef and lamb, as are the Germans and French, who, incidentally, are also indignant at the thought of Cadbury's chocolate being labelled as chocolate. Add it all together and you have a multi-neurotic community, obsessed with small regulations.

Rules about insecticide residues or the shape of cucumbers. Rules about the cooking of food in restaurants, the killing of animals in slaughterhouses or the sale of national cheese. Rules about the bumpers of cars. Rules about the sale of beer. But these rules are often rooted in national culture. Standardising rules on ten thousand matters of life is impossible without standardising culture – or destroying it.

Hamburgers in the US are now subject to 41,000 regulations involve 200 laws and 110,000 precedent-setting court cases. They range from the thiamine content of a bun to the thickness of ketchup and the level of pesticide in meat. Europe is going the same way.

Take the culture clash between Greek and German culture. Holidaymakers in Corfu know the relaxed attitude of the Greeks to almost everything in life, including whether or not motorcyclists actually wear helmets. While the law says they must, the Corfiot roads are clogged by riders in swimming costumes zipping along, hair flapping. We can't do this and we can't do that. Why not? We've always done it like this before. Because the EU regulations say we can't – any more. But laws make life more difficult and expensive for manufacturers, distributors and retailers. The cost of living rises, salaries rise, the cost of production rises, jobs fall – moving to countries where the bite of law is less fierce.

Strains in Europe from monetary union

The strains in Europe will be immense. Monetary union means that the foundations of economic policy are now dictated in most countries by a majority of others who do not even speak their own language, and perhaps have never lived in their country. But monetary vision without progressive alignment on all major issues will be impossible. Common immigration, asylum and visa rules, common defence (or attack) policies, common foreign policy – all these things to agree on or be steamrollered over. Then there are the legal strains.

So what will happen to Europe?

The core nations of the old EC will continue to move together to align their economies and many other areas of life, in a process which will continue until the day the whole thing starts to fall apart. No nation will want to risk being left out, even if it means economic gridlock. Yet those same nations will in the future be just as relieved to get out of the all-embracing European mega-politik.

Pressures will remain to protect national economies from currency instabilities through regional economic alliances, and these pressures will increase as further globalisation produces ever larger surges on the foreign exchange markets. The way ahead will therefore be very difficult and messy.

The second-level nations, such as Poland and Romania, will be struggling to get in but will later consider themselves perhaps fortunate to have been kept out during the first difficult unification process. The arrival of 10 new nations at the heart of Europe will impact areas such as common military and tax policies.

European Monetary Union

And what of the single Euro currency? One thing is clear: it won't be sustainable without pain. The individual countries are in too much of a mess for that. It will continue as a fudge, with France, Germany and other countries squeezed into a straitjacket from which there is no escape. The harder the squeeze to get in, the greater the agony of staying in. Countries with different economic problems and in a different stage of their business cycles will experience great pressures. Imposed squeezes on budget deficits will create agonies for public spending on health, education and social security – all three sensitive areas for the electorate.

Never again will a member country be able to allow its currency to devalue to offset inflationary pressures, or to set interest rates at a higher level than its neighbours to control inflation. All the Euro-zone members will swim or drown together. But the conditions that enable some to swim will drown others.

Many businesses are keen. They cannot understand how nations in Europe could have hung on to their own currencies in a globalised world which demands currency stability. Unilever is just one multi-national threatening to pull out of Britain unless it joins the Euro. For them dealing in 15 currencies used to be a nightmare.

So there will be a conflict between 'corporate' tribes who want a simple trading area across the region, and 'people' tribes who will sometimes be very hostile to the emerging meta-state.

The single currency will have a dramatic impact on some businesses, and will benefit all consumers over the next decade. Car retailing is a lottery at present: there has been a 30 per cent difference in the price for which the same vehicle could be bought in different parts of the EU. Slicing up the EU into cosy national markets will have to stop. Pricing has been rigged according to what local people will agree to pay. Cross-border shopping on price will then be dictated by tax differences, which are also likely to converge.

One of the destructive pressures on the 'United States of Europe' will be high unemployment caused by changing conditions and labour force immobility inside Europe, combined with a progressive, long-term shift of manufacturing and service provision to the lower cost Pacific Rim economies and to other countries such as India and, later, African nations. By 1997 Germany already had more people out of work than at any time since Hitler came to power, with dole queues four million long. This problem will partly be offset by labour shortages in some areas caused by an ageing population but has so far grown worse.

For individuals, this all adds up to a difficult, rapidly changing labour market where those with rare leadership and communication skills command vast salaries, while manual labourers are semi-permanently unemployed. Senior executives will dominate a global stage, but the next layer down will often be under pressure to relocate east, to manage subsidiaries in Asia.

One concludes that unless there is a radical rethink on policy, all Europe will have left to offer by 2035 will be a massive cultural museum (history for tourism) and the intellectual capital of its age-ing workforce which is relatively immobile (for example, software development skills and management consultancy). However Europe will benefit in the short to medium term from relative instability elsewhere, becoming a haven for investors. Dictatorship-led econo-mies are inherently unstable due to the lack of openness, account-ability and transparency in their governments. This has been a factor among many in the currency crises in Asia and elsewhere. Dictatorships also have a habit of being overthrown, followed by further internal conflicts and uncertainty.

THE EMERGENCE OF THE ENGLISH TRIBE

Anti-imperialism (a reaction against control by one country of another) is tribalism. Anti-imperialism is so strong as a global force that any kind of new world order will be extremely difficult to establish without force, unless as a creeping bureaucracy of inter-national regulations, or response to common threat (see Chapter 6). Any hint of imposed rule by one tribal group over another will continue to be fiercely resisted. That is why the Scottish nation has been so obsessed with the thought of being free from Westminster rule 'by the English'. It is the reason why for generations the British parliament has over-compensated, by allowing the Scottish elector-ate far more MPs than by rights they would otherwise have been allocated given the small size of their population.

The English have suffered an identity crisis of their own. Historically they were strongly globalised and strongly tribal. The Empire was the centre of power on earth and the Union Jack a potent symbol of global supremacy. 'Britannia rules the waves.' Yet England itself has very little native culture left – apart from the pomp and ceremony associated with royalty.

Royal reforms and a national flag flying

There is, of course, the Royal Family, with a dearly loved Queen and Queen Mother, but with a set of children and their partners (or ex-partners) whose antics sell papers but who have lost the moral capacity to lead by example. The British people will not stomach a Grand Monarchy in the third millennium headed by King Charles III after all the troubled events involving his former wife and mistress. The trend to a 'welfare monarchy' will continue. There will be less pomp and more social work, royals fund-raising for good causes and encouraging business to help create a more caring society.

The tragic death of Princess Diana will continue to overshadow the monarchy throughout the young adult years of Prince Harry and Prince William. The Royal Family will never be the same again, experiencing huge pressures to complete a reform process, become less formal, less distant and less expensive. Nevertheless tribalism will save the monarchy, precisely because there would be so little left otherwise to make us British. The fundamental problem is that by definition royalty is based on genetic discrimination: unless you have the right combination of royal genes, you cannot fill certain roles. This genetic elitism will seem increasingly bizarre and morally suspect to a people who have fought for equality of opportunity, fairness, and lack of discrimination.

There is a class system in Britain which is as destructive in some ways as the caste system in India. There is a ruling class, which by virtue of its family line has enjoyed genetic rights to sit in the House of Lords forever. Royalty is part of this system. I am a monarchist, not a republican. I think the British people will be the poorer without a constitutional monarchy – but expect radical reforms by 2010, with major cultural changes already obvious by 2005.

Then there is the national flag – but of what nation? Scotland can fly a national flag but those in England will probably fly the Union Jack. A few might fly the English flag of St George, but confusion and guilt over the Empire meant that from almost the last day of the Second World War until the anniversary of VE day some 50 years later, there was hardly a Union Jack or a flag of St George to be seen fluttering from a public or private building in the whole of England. Most Union Jacks in Britain were printed on cheap T-shirts and plastic hats for tourists.

As the UK continues to disintegrate in the final death pangs of the English imperialistic dream, you will see a rebirth of the English people: a fresh energy in a new generation who want to find ways to express that they are as English as the Scots are Scottish or the French are French. National state funerals, the last night of the BBC Prom concerts, international football matches and other events will help focus national identity amongst a people increasingly feeling the need to be proud to be English.

The M generation are tribalists: 88 per cent of teenagers would choose England to live in, 66 per cent think of themselves as English not British, 72 per cent say nationality is important to them, and most teenagers expect the UK to divide into separate states with their own border controls and passports in 20 years' time. While the Scots may look to Europe as a way of staying together in a broader alliance, preferring to accept some rule from Brussels rather than close rule from London, the English are likely to become increasingly resentful of the new non-elected emperor, sitting somewhere in Europe, presiding over a strengthened but distant and distrusted European Parliament.

TRIBALISM IN ASIA

The threat to China

China is reforming at a remarkable rate, embracing many aspects of the market economy with enthusiasm and success. But tribalism could one day destroy the cohesive might of China. This ancient superstate has 1.2 billion people, more than 48 languages and many ethnic minorities. While over 90 per cent of the population are

Han Chinese 'sons of the Yellow Emperor', large areas are inhabited mainly by other groups.

An example of recent pressures is in Xinjiang. Since a major uprising in the border area in 1962 there has hardly been a year without trouble there. The local population of Turkish ethnic minorities outnumber Han Chinese by two to one. Kazakhs, Tajiks, Kyrgyz and Uighurs are seeking to rebuild ethnic and nationalist ties with the new central Asian republics. There have recently been pro-independence Muslim riots (there are 20 million Muslims in China). The main separatist group based in Kazakhstan says 57,000 were arrested in 1996 and 100 Islamic schools were closed.

Such separatist tendencies are strongest in Tibet. The Chinese government has used its usual technique of control by flooding Han Chinese settlers into Tibet's towns, but with little effect. Even in Inner Mongolia where Han Chinese outnumber Mongols by six to one, Mongolian nationalism threatens to flare up at any time.

Tiananmen Square – and the simmering pressure cooker

The history of dictatorships is that eventually freedom breaks out. What about China? The crackdown on open political debate there is unsustainable. It means that 1.2 billion people cannot express what they feel except in secret. After terrible mistakes and famine under Mao Tse-Tung, Deng Hsiao-Ping began to open China again in the late 1970s. In the 1990s China tried in a few years to do what the West did over centuries. The Communist Party was swept to power by violent, explosive turmoil. Many of the leaders in China in the early 2000s lived through the Cultural Revolution of 1966–76. They know the power of angry people.

The protests in Tiananmen Square were against official corruption but corruption remains and secrecy, as seen in the early responses to SARS infection. One day there could be another revolution, a risk of civil war. The alternative is that China reforms, embracing the market and also other freedoms that those in the old Soviet Union are beginning to see. There, the country is loosening up, despite President Putin's centralist tendencies. Foreign TV can be received in many areas by satellite. Internet access is uncontrolled when someone dials an international call.

There is another rapid cultural revolution taking place, which is almost invisible but profoundly life-changing for the nation. China's present leaders went to study in Moscow, but their children, with thousands of others, go to Yale, Chicago or the Harvard Business School. Combine this with huge technical assistance in-country, whether from the World Bank, the International Monetary Fund or multinational companies, and the influence at senior levels is profound. The successful transformation of Russia into a strong, vibrant, secure country will also do much to ease reforms in China.

China can remain united and strong, but only by providing continuous rapid economic growth. That can only be achieved by allowing Shanghai-type, large-scale inward investment and ownership, in many other parts of the country. But the invasion brings contact with the West, net and TV as well as material wealth. Expect huge pressures for further liberalisation of the people, freedom of expression.

Japan facing problems

Japan will itself face severe problems from an ageing population and cultural isolation, as well as from a mentality that does not encourage creativity. Japan produces 10 per cent of the world's economic output. In 1992 it produced more cars than any other nation, 15 per cent of the world's steel, launched more ships than anyone else, produced more televisions and radios than Europe, more watches than Switzerland and was a global player in the computer, aircraft and space industries.

Expect Japan to have difficulty maintaining its economic might, surrounded by emerging but vulnerable economies with cheaper labour and high technology. Expect trade restrictions to ease, allowing Japan's high trade surplus to settle. Expect Japan to own big 'foreign' industries and parts of cities abroad, and to make far less at home. China will affect Japan's future.

Expect Japan's consumers to go on worrying about buying goods in a deflating economy, with stagnation in domestic markets and low morale affecting the country at least until 2005.

Tribalism in India

India is the greatest democracy in the world, yet few countries are so tribally based, with its deeply rooted caste system. Tribalism will threaten to break India up, for it is more a continent than a country. With over a billion people, a single state has more than three times as many inhabitants as the whole of the European Community. A flight from one end of India to another is equivalent to flying from London to Moscow in terms of millions of people flown over and language groups passed by. India already shows signs of fragmenting.

To the far east, beyond Bangladesh, close to the Burmese border, are the north eastern states such as Manipur. These have been run almost in a state of emergency recently due to tribal fighting and strong independence movements. More soldiers died in the north east in 1997 than in Kashmir. Ethnic groups here are very different in every way from the rest of India. In facial appearance they almost seem Chinese. But India will survive despite recent religious tensions.

So this is another paradox. People want to be part of a globalised planet and to function in free trade areas with freedom of goods, services and people. Yet most groups of people are still fiercely territorial, and are becoming more so.

ETHNIC CLEANSING

In the meantime, tribalism expressed through the horrors of ethnic cleansing will continue to haunt Europe, its memories still raw from the Second World War and now from Bosnia and Kosovo.

There is nothing new about ethnic cleansing. The deportation of unruly minority groups has been a common practice of victorious armies since ancient times. After the end of the Second World War the allies allowed the Czechoslovakian and Polish authorities to expel seven million ethnic Germans. But the results are often horrific. One million dead in 1947 during the partition of British India, over half a million dead in Rwanda in the mid 1990s, tens of thousands butchered in Bosnia. The complete exodus of the Asian community from Idi Amin's Uganda in the 1970s was less bloody but no less sudden or dramatic.

Expect to see more ethnic disputes as mobility and immigration muddles up the original racial mix of nations, transforming cities and rural areas. Tribalism will produce troubles which have no simple solution, with terror and bloodshed followed by yet more refugee movements.

Tribal minorities demand compensation

In Australia a court judgement has raised the spectre that almost 80 per cent of the country could be subject to legal claims by Aborigines. The High Court ruled that the Wik people had a valid claim to land leased from them in 1915 by white farmers. Already 40 per cent of the country has been subject to native claims, including the bulk of western and southern Australia. Relationships with the Aborigine population have not been helped by the discovery that thousands of children were removed compulsorily and sent abroad in previous decades. This scandal has yet to be sorted out. Expect similar disputes to continue in many other indigenous groups.

Population pressures will be used to justify ethnic massacres

As we have seen, 98 per cent of the growth in world population is in developing countries. Expect to see a surge in the population of China as its dictatorship finally collapses at some stage early in the next century, releasing a huge pent-up demand to be able to have children freely.

'There are too many people on the earth'

Since the late 1980s I have heard many comments that the AIDS disaster in Africa hardly matters because it is a way of 'controlling the population'. These shocking comments have come from the well educated and less well educated, from influential 'movers and shakers' and the poorest in other continents. More recently I have heard the same comments in a chilling response to the rapidly worsening AIDS situation in Mumbai and Calcutta: an indifference to plague because people think it is nature's way of thinning out the unwanted and unneeded. This is particularly the case in India, where at present many of those dying with AIDS are those in the lowest ranks of society.

Expect the same arguments to be used by undisciplined armed mobs as they target unpopular ethnic minorities. Expect to see the 'culling' of human populations by horrific acts of slaughter, justified by the claim that 'there are too many people on the earth'. The same idea will be directed at the old, infirm, sad, marginalised and all those who at the least pressure drop off the edge of mainstream society. Expect more mass graves, and mixed feelings in some at the news of massive loss of life in densely populated poor countries from flood, avalanche, disaster or plague.

The larger the world population, the more spectacular the scale of human disasters as more and more people are packed into more and more hazardous areas. The low-lying landscape of Bangladesh is a prime example: permanently at risk of catastrophic flood as a result of annual monsoon rains, a risk made worse by recent changes in the use of land. There are 116 million people, a figure increasing by 2.6 per cent a year, who live at the convergence of three great rivers: the Ganges, the Brahmaputra and the Meghna. Much of the land is less than 15 metres above sea level. Rising population densities and rising sea levels caused by global warming are dual risk factors pointing to a vast human disaster in the region by 2025. But tribalism will ensure that few of these millions are welcomed into other nations.

Territory is at the root of tribalism and national identity

When increasing numbers of visitors become residents, then nationals, then voters, then rulers, resentment is likely to occur at a certain point. Immigration laws and laws against racial harassment are almost impotent against minority extremist groups who take the law into their own hands. You cannot guard the front door of a vulnerable family on a tough estate every hour of every day. You cannot prevent all the cruel taunts at school, all the petty bullying, the catcalls from those in the street, the mindless verbal abuse, graffiti and stones thrown at windows. These are the realities, and they are one reason why ghettos form.

Here are trend and countertrend: racial mixing and ghettoisation. Or are they both part of the same thing? Who do I belong to? Do I belong to a nation that I was not born in, or to an ethnic or cultural group within it? Ghettos will continue to flourish in big cities as relatively safe, geographical culture centres for different groups, whether identified by ethnicity or by lifestyle.

TRIBALISM AND GLOBAL BUSINESS

Global citizens

Expect to see a new breed of hyper-mobile globalised people with no national loyalty or identity and no commitment to any geographical area, yet with friends in every city. These industrialised techno-gypsies consider themselves global citizens. They will be hard to tax and hard to count in census surveys, as well as hard to police.

We see them already: for example the executive who spends more than half the year away from home, equally divided between two other continents, and who decides to take a flat in another city, which he visits frequently. Where is home?

Tribal conflicts can brew up fast

The start and end of major tribal conflicts can be hard to predict and may affect corporate investment. A huge mining company like Iscor in South Africa is faced with multiple ethnic challenges. In the light of recent tribal warfare, should it risk a new six-year commitment in central Africa to develop a new mine from scratch, including infrastructure investment such as building a road to the mine? What will the current instability be like then? Is it sensible to think in such a long time frame? Many corporations are turning to short-term, in-out projects in emerging countries where they fear loss of longer-term assets. They want to be able to make their profits fast, cut their losses and run.

Pacific Rim pushes west

After the huge growth of the Pacific Rim economies, what next? Wages will rise as these economies heat up again following major setbacks in the late 1990s, with local shortages of skilled labour and increased expectations for living standards. It will take several decades for the process to have its greatest impact on China. India and neighbouring countries will gradually get swept up in it all. After all this there will be only one part of the world left with the best part of a billion people available for ultra-cheap labour – sub-Saharan Africa. However, the automation of most production worldwide will marginalise them, too far away to be employed in service industries. Aggravated by AIDS, poverty in Africa is likely to remain one of the world's greatest moral challenges.

Southern Africa growth engine

South Africa and Zimbabwe should have had the wealth, expertise, networks and infrastructure to help accelerate the rest of Africa into the third millennium. Despite recent problems, there are many entrepreneurs in southern Africa who have grown their bases and are looking north, capitalising on rising costs of labour in the Far East. Many of these young companies will experience spectacular growth. Import/export, manufacturing and a host of other new industries will flourish, working against severe effects of AIDS and local instabilities. The Southern region will not recover until the end of President Mugabe's disastrous rule of Zimbabwe.

Does the lift factor actually work?

What will happen to the poorest of the poor? Will they be lifted as these national economies grow, or will they be trampled? Some have argued from experience in Indian cities such as Mumbai which have experienced a boom, that the rich get richer while the poor starve and die. That may be true in the short term but not in the medium to longer term, so long as developing countries develop a social conscience alongside their increasing national income. This is likely to happen in the longer term as part of the process of westernisation, which continues to export Americanised Judaeo-

Christian values about human rights and worth. The larger the economy, the greater the state revenue and the greater the choices about redistribution of some of that wealth to protect the most vulnerable.

Logic dictates that the trickle-down effect will help relieve absolute poverty while not dealing with increasing inequality. How else can the extremely wealthy spend their income except by paying others for their time or their property? Either way, money begins to circulate in higher volumes than before. Productivity rises as the unemployed or only partly employed have the chance to work hard, increasing the standard of living of the community. And those who are already employed become steadily more efficient with new tools, technology and education.

The greater the contrasts between rich and poor in a society, the greater the risk of instability as antagonism, aggression and organised opposition grow. Therefore healthy societies will always tend to redistribute wealth to some degree, even if in the interests of self-preservation rather than because of a troubled conscience.

Despite this, expect gaps in gross pay to widen, with even higher remuneration packages at the very top for outstanding leaders and advocates, people who make things happen. One reason why such large packages will be needed is to entice such people into helping companies perform, since all the most successful will by definition have no real economic need to work any more. Despite the huge financial rewards, many senior executives will be driven mainly by the challenge of the task.

Another reason for big payouts will be the severe scarcity of world-class leadership, communication skills and creative genius. If a multi-billion turnover company plunges into the red following two disastrous CEO appointments, it is in the interests of the company to pay a huge price to guarantee sorting the problem out fast. Payment by results is the easiest way to get the biggest deals approved, and this will continue to be common. But how do you measure results? Pay packages based on balance sheets often encourage the wrong decisions in the longer term. Expect to see more sophisticated measures, such as 360-degree assessments.

At the other end of the scale, expect micro-banking income generation projects to grow spectacularly in many deprived megacities, each lifting tens of thousands out of absolute poverty in less than twenty years, creating a new generation of middle class business owners. The results in Indian slums are spectacular.

Tribalism means respecting culture

All the globalisation in the world will not produce identikit nations. Indeed, national identity will become even more important. There are too many differences, for example, for global TV to be run by robots, offering the same channels everywhere all the time – as a French technician found out recently to his cost. His pulling of the wrong switch resulted in children's TV being replaced in Saudi Arabia by hard-core pornography from Club Prive de Portugal. To add to the insult, the programmes were beamed via a Saudi-controlled satellite operator, ArabSat, which was in partnership until that moment with Canal France International, a state-financed French television company. That was the end of a sweet relationship. The companies that succeed in a globalised world will be those that understand local culture, for example in management styles.

TRIBALISM IN RETAILING

Tribalism is a dominant force in retailing: people have intense brand and retailer loyalty. Designer clothes thrive on tribalism. Brands like Nike will continue to charge big premiums for the privilege of wearing a label, particularly among teenagers and young adults where labels will continue to attract cult followings. Brand name forgery will continue to grow. Buying the label is buying a place in the family. The more upmarket the customer the more tribal they tend to be. Customer loyalty means that 'I buy all my food at Marks and Spencers and all my clothes from Harrods.'

Fashion itself is built on tribalism. A style is by definition a means of identifying with a group. There is no such thing as individual style. The style of a single person is not a style, just an expression of eccentricity. Style is a common theme adopted by many.

Niche marketing is usually tribal

People who occupy a particular niche are likely to have a host of things in common and successful marketing will use every one of those factors to hit home. And marketing forms the image of the tribe. It also plays on and exploits the tribe.

Product tribalism

In a world where people crave secure relationships it is a shock that relationship marketing is such a new science. Brands can act as relationship partners for consumers. Interviews with hundreds of consumers have created a picture of 15 different relationships that consumers have, in which they 'belong' to their products or the other way round. These range from the long-term, committed partnership an athlete has with a make of trainers he believes helps him win, to a childhood friendship when someone buys food that they used to eat as a kid, to a secret affair when someone sneaks down at night to raid the freezer for a hidden tub of luxury ice cream. Expect to hear much more about this. Indeed 'relational' or 'relationship' has become the latest 'in' concept to attach to just about everything. Relationship marketing equals 'fostering a special alliance with customers by gathering and employing massive information about individual behaviours and buying habits.'

Hundreds of corporations have already jumped on the relationship bandwagon. Their Heads of Relationship Marketing and Relationship Marketing Departments are advised by Relationship Marketing Consultants. In part the industry is being built out of information tools only recently available. A key target is direct marketing. Direct marketing is a major specialism with a proliferation of new courses and qualifications to generate the high-calibre people who will be needed to sustain current spectacular growth.

Toyota in the US created a relationship marketing programme aimed at students. The goal was to use the Internet to capture 5,000 names for a database of college-age potential buyers and sell them 550 cars. The X-treme Fun College Incentive Program was rolled out in mid 1996 with heavy advertising for a multimedia website. In fifteen months they had 33,000 names and had sold 21,178 cars. 'We are building a relationship with college students as they age,' they said. Respondents to their ads were offered a $500 rebate and a two-year warranty. Forty per cent came via an 0800 number, the same percentage on the web. Details of parents were also gathered for targeting in a second wave.

But relationship marketing can backfire. A major supermarket chain sends a mailing about disposable nappies to a man, addressed as 'Dear shopper, being a mum can be such fun!' There can be no more effective way to advertise that you have no relationship than to make such an error. Expect, after all the hype, that relationship marketing will come to be seen as little more than what companies always did: care for existing customers and target marketing to specific groups – but using better information systems.

Tribalism in sport

Tribalism will continue to drive sport as much as the love of sport itself: a supporter of a local team for example. Take tribalism away and sport falls to pieces as a spectator industry, becoming just a collection of individuals trying to excel. Even single-person competitions such as golf create tribal followings. Successful football or baseball clubs are only as successful as their tribalism. Tribalism brings in crowds, sells merchandise, attracts sponsorship.

Advertisers love tribes

Tribal gatherings are an advertiser's dream. Create a tribe and money follows. So the organisers of a conference of doctors know that they can bank on massive subsidies from drug companies, who pay for stands and other promotions. Create the World Economic Forum for global leaders in industry and large-scale company sponsorship begins to arrive. Start a new radio station with a very narrow (high-income) target group and you can set the budget high. Forming, manipulating and exploiting tribes will all be key functions of many successful organisations beyond 2010.

Expect global branding to be questioned where the brand is closely linked to one nation, culture or religion. Expect many global brands to hide inside local 'tribal' packaging.

TRIBALISM AND THE MEDIA

Tribalism will have a huge effect on news reporting in the media over the next decade, especially on TV. Tribalism means that you identify with your own kind and are less interested in news from elsewhere. Expect ratings to continue to fall for pre-millennial style news bulletins on TV. Slayings, beatings, spectacular suicides and disasters, together with pictures of starving millions and other tragedies will be turn-offs unless they are relatively local. There is not enough real local news to keep ratings high. Most days are fairly boring and repetitive. CNN's daily audience in the US in the second quarter of 1997 fell to the lowest ever in its 17-year history with a mere 284,000 viewers. If each watched a whole hour of news a day, then the average US CNN audience for any given bulletin during the three months was little more than 10,000 people – maybe far fewer. Wars boost ratings – but the underlying trend remains.

Fox News has barely 25,000 viewers a day, or an average of 1,000 in an hour. MSNBC scores just 40,000 a day. These stations are hardly going to change America. They are competing in a tiny and shrinking market. Expect CNN to alter its format and remain profitable with advertisers attracted by an elite and interesting audience. In the US alone, advertising raised around $550 million a year in the late 1990s, double the figure in 1990. The attraction is a global audience (which is impossible to measure). In theory 500 million people in 210 nations live in households where they have access to CNN.

Expect global turn-ons for all major world events which capture the imagination. When Princess Diana died in 1997, her funeral was watched by 2.5 billion people, out of a total world population then of 5.7 billion. CNN scored highly, with 1.8 million American viewers on the day of the funeral. It was a prime example of technology and globalisation – two and a half billion people simultaneously participating in a live act of collective grief. It was a defining moment in British life. Nevertheless such events will not change the underlying trend, which will be towards the tribalisation of news.

News gathering has yet to catch up with new technology. In a global village there is so much news, but how much of it is relevant? Most world news is nothing of the sort, but is rather a US-dominated rehash of the day's events. Expect CNN, Fox and other stations to try to lose their US 'feel' so as to compete better in the global arena.

When you have a population of 6 billion people, every day the law of averages dictates that somewhere in the world there will be an aircraft crash, a rail crash, a terrible car pile-up on a fast road, a terrible industrial accident or an appallingly violent act of terrorism. But what makes the news? News is driven by images that are available. Sensational footage of a train plunging down a ravine will guarantee that the story makes top news worldwide, regardless of location or how many killed.

But news driven by the sensational will always tend to be news of the dreadful, and an unlimited diet of the dreadful is depressing – the opposite of entertainment. Expect much soul-searching over news in tomorrow's world. News as entertainment and as tribal gossip – the ratings war will do it all.

Newspapers work better than TV for news because they contain more stories (a news bulletin will cover only four or five), and that allows selection. Expect TV news to hit back aggressively with interactive news, using the Internet and digital TV. Video newspapers. CNN Interactive, ABCNEWS.com, MSNBC and Fox News all streamed live video coverage of Princess Diana's funeral online. CNN recorded 4.3 million page hits on the day of her death. ABC News recorded 2.9 million hits in the first 48 hours. CNN made many film clips available online that never made their way into normal bulletins. Internet news has finally come of age, changing the future face of newscasting forever. And of course news on the Net is always likely to be more complete. In the case of the death of Princess Diana, for example, various sites carried pictures, taken by the paparazzi, of the Princess in the crash car.

Expect news gatherers of the future to be members of the public equipped with nothing other than a video phone, able to be eyes and ears, transmitting live pictures and excited commentary around the world, seconds after phoning CNN or the BBC, from their location in the thick of major events.

POSITIVE TRIBALISM

Many of the earlier examples of tribalism have been negative, to do with nationalism, racism, elitism and sectarianism, yet tribalism is an immensely positive force. Tribalism is the basis of all family, team and belonging. Tribalism provides a sense of identity. Tribalism helps us understand who we are, where we've come from and where we're headed. Teams are to do with tasks, tribes are about whole groups moving together.

Tribes hold the whole of society in a common community. Neighbourhoods are tribes, members of sports clubs are tribes, football supporters are tribes. If there were no tribes, human beings would create them in a day. We need tribes to exist, to make sense of our world.

Tribes are like countries on a map. Without tribes there is no geography in our relationships because there are no groups, just atomised collections of isolated individuals relating equally to everyone. Therefore if you want to understand the forces on someone's life, their motivation, the basis of their values and decisions, you need first to understand the person's own tribal culture.

Of course these cultural factors are particularly obvious when working in a multinational corporation, with executives from very different countries, but they are often no less significant in other, more localised working situations.

Corporate tribalism

Tribalism is a very powerful concept for large organisations. Every organisation is a tribe and there can be many tribes inside corporations: front and back office, HQ and regions, sales and credit control, research and marketing. Tribalism in a company makes us proud to belong. Tribalism can weld teams together in a very healthy and competitive way. Corporate tribalism raises key issues: can we have one culture for an international company? What about cultural adaptation? What about corporate discipline?

Tribalism is one of the most powerful tools managers have for increasing productivity, competitiveness and company loyalty.

Every manager needs to understand how to create and keep a tribe, and how to belong to a larger one.

Building a tribe is more than team building. Teams are groups with common goals and functions. By definition there are severe practical limits to the size of an effective team. But tribes are different: pro-active, competitive, proud-to-belong groups which can form and re-form with their own culture, identity, loyalties and driving force. Tribes embrace everyone in the workforce. Developing a successful tribe depends on inspirational, dynamic, visionary leadership. The greater the leader, the larger the tribe can be.

Tribes are as large as whole departments or entire organisations. Successful tribe formation in companies requires a shift in management style. It means the promotion of culture and feel-good identity as well as activity. It means vision building and communication.

The fastest way to change an organisation is to address the tribal structure, which is far more than mere team building. Take a multinational: if Frankfurt HQ wants the New York office to change, it needs to target the Tribal Chiefs in New York. Who really commands respect? Replacing a New York Chief with someone from the head office in Germany may backfire. Team logic might suggest it's a good idea to bring HQ priorities and culture right into the operation, especially if the remaining senior team work well with the new Chief. But the rest of the tribe further down may be far from happy. We could be talking about powerful negative perceptions throughout the entire US operation. And those negative perceptions will directly hinder change as well as damaging morale, productivity and bottom-line profit. Smart managers take care of their tribes, understand them, work with them and harness their strength.

Tribes don't form in companies overnight. You can create a team in a week but a tribe comes into being. Tribes form all on their own and can be the basis of pressure groups within an organisation. You can work with tribes or you can find they work against you. You can accelerate their development but it is hard to cause their destruction – without mass redundancies and relocations.

At a time of constant change, a real challenge will be to preserve the vitality of an effective tribe. Take two large companies in a merger. Are both to lose their unique identity? Is one culture going to be dominant? Mergers often involve large-scale redundancies. If tribal identity is also lost for those who remain the result is a double demotivation. The process requires strong leadership, pulling two tribes together.

Tribes in companies make money

Research shows that creating a sense of family at work increases productivity. Friendship between employees help significantly. A recent study found that groups of friends produced three times more manufacturing output than acquaintances and 50 per cent better decision making. They benefited from increased trust, honesty, open communication and respect.

Family business

Expect most new wealth to be created (as at present) by small companies with fewer than 20 employees, of which 75 per cent or more are family owned or controlled, and have been started with family money (or with loans against equity such as the home). Expect these family-based firms to resist employment legislation such as equal opportunities, quotas and other restrictions. They will continue, despite legal challenges, to discriminate widely in favour of relatives, friends and friends of friends as future employees.

Expect new government initiatives to help provide longer-term venture capital for smaller businesses, driven increasingly by pressures from commercial lenders for near-instant results.

CHALLENGES TO MANAGEMENT

Tribalism as an issue
- Has your company ever looked at tribalism as an issue – national culture, target groups, markets and corporate tribalism?

Language identity

◆ What is company policy regarding national languages and company language?

◆ Does this need reviewing in the light of current trends?

Culture sensitivity

◆ What kind of preparation do you give staff when posted cross-nationally?

◆ How sensitive is head office policy to local cultural issues?

◆ Is it sensitive enough?

◆ Is your company able to capitalise on new tribal forces – for example emphasising the American-ness or Frenchness of a product?

European instability

◆ Is your company prepared to exploit economic union but with increased inter-country tensions, protests, marches and other popular disruptive action?

Ethnic cleansing

◆ How effective are your anti-discriminatory policies?

◆ How possible is it for someone with very negative views of another people group to pursue departmental actions which discriminate, e.g. in employment?

◆ How vulnerable is your company to charges of racist actions?

Tribal marketing

◆ Has your company invested enough in relationship management, building a tribe of loyal happy customers?

◆ What about relationship marketing?

Tribal team building

◆ How strong is the corporate 'pride factor' among your workforce?

◆ What steps can you take to increase positive tribalism inside the company, including competition between strong teams?

PERSONAL CHALLENGES

What tribes do you belong to?
- Where do you belong?
- What groups do you identify with?
- Where do you get your inspiration from?
- Are you happy about the tribes you are identified with?
- How do people label you?
- Is there an exclusive tribe you want to belong to?
- How could you join?

Changes in Europe
- Are you prepared for major changes in Europe?
- Do you have contingencies in place both for the success of further integration, and for problems and conflicts?
- How would these things affect you personally?
- How European in thinking are you?

Racial discrimination
- Are you aware of racial discrimination in your workplace, and what are you doing about it?

Tribalism as a management tool
- After all the management emphasis on teams, have you considered working in tribes, harnessing these human forces to create larger dynamic, competitive groups with a common sense, identity and purpose?

Cross-cultural understanding
- How good are you at relating to people of very different cultural backgrounds?
- What are you doing to increase your understanding of different cultures and ways of doing things?

CHAPTER 4

Universal

Global management

The fourth face of the future is the exact opposite of tribalism, and is universalism: McDonalds everywhere. The greater the globalisation, the greater the tribalism. One accentuates the other. Tribalism and universalism feed each other, each the reaction to its opposite. The greater the global uniformity, the greater the drive to maintain tribal identity. The more secure people are in their own identity, the less they are threatened by globalised sameness.

Globalisation has been a long-standing feature of the media. American culture has been exported on a large scale since before the Second World War, mainly through films, TV and popular music. Nations have cultural power, or soft power, that is greater today than the hard power of military strength, through the influence of culture, values and the perception of a technologically superior society. As China has recently recognised, in world power stakes nuclear weapons matter less in influencing international affairs than the promotion of global values in the mass media.

Globalisation is an unstoppable force. Even President Jiang Zemin accepted it as a recognised fact in his keynote address to the 15th Chinese Communist Party Congress. Globalisation is the result of the free movement of capital, goods and services across national boundaries. Globalisation has increased competition and lowered profit margins in many countries. Globalisation is the result of decisions by the governments of more than 170 nations to operate more and more as part of one global unit. Where does it lead us? No currency flow restrictions. No import or export tariffs. No national protection against sudden fluctuations in external market forces. Greater freedom for capital movements which result in severe, destabilising currency volatility. And at the opposite extreme Burma (Myanmar) and North Korea, which have been impoverished by their extreme isolationism.

A WHOLE WORLD TRADING TOGETHER

For the first time in history almost the entire population lives in a global capitalist system. The driving force towards globalisation is economic growth and prosperity, especially for poorer nations whose economies have often been the most restrictive in the past. They have been propelled by statements such as these from the World Bank: 'There is a positive link between freeing markets and trade and the eradication of poverty in the long term' and 'There is no evidence to justify fears that free trade pushes down wages for unskilled workers in developing countries.'

But the UN has a different view. 'Increased global competition does not automatically bring faster growth and development . . . In almost all developing countries that have undertaken rapid trade liberalisation, unemployment has increased and wages have fallen for the unskilled.'

There is a fundamental problem with globalisation which will cause international tension and trade disputes without arresting the process. The problem is the irrational nature of the global market, coupled with the extreme vulnerability of the poorest and most marginalised in emerging economies to sudden changes in exchange rates, interest rates, or big investment decisions.

Consider the following scenario: Country A has a rapidly growing economy. Many companies are booming. Foreign investment is pouring in. Property prices are soaring. Businesses are borrowing ever larger sums, with little or no security except their expectation of future large profits. Every month these companies have to borrow more to buy more stock to make more goods for ever larger orders, which will be paid for in the future (they hope there will be no bad debts). They are also exposed through large assets held in property. There is little inertia in the economy. Currency reserves are tiny compared with hourly currency flows by global institutions.

Then comes one piece of unsettling news and currency selling begins. Traders may be confident that the currency is now under-valued, but they will go on selling as long as they believe other traders think the currency is still overvalued. In other words, buying and selling becomes driven not by objective data, but by what they think others will do. But this is a recipe for overshooting, seen over and over again in currency, commodity and stock markets.

A bizarre situation can exist in which everyone privately thinks that the currency is already too low, but continues to sell hard only because they are certain that everyone else thinks the currency still has further to fall. Rates fall through the floor in a mass wave of panic selling, as dealers dump currency in the near-certain know-ledge that they can buy it back at a profit in a few minutes, hours or days. The big issue is not what the real value of the currency should be in the light of the economy, but how the rest of the market is likely to behave in the very short term.

Free market dogma is that these peaks and flows will always sort themselves out. 'Don't try to buck the market.' True as this may be, it ignores the monumental impact of these arbitrary swings on families, communities and nations.

The big difference between Britain, the US and, for example, Thailand, is that the ex-workers of a bankrupt company can often still eat, drink and have homes in the West. In Thailand there are very few safety nets. If you have no job and are already poor you don't eat, your family gets little or no health care. A massive fall in currency may last only a few weeks before partially correcting itself, but that is plenty long enough for multiple bankruptcies. Some companies can't afford to buy the foreign components they need for manufacture. Others are crippled by sudden increases in interest rates to support the currency. Yet others collapse because a large creditor is suddenly unable to pay a bill. Banks fold as companies default on repayments and as property prices fall below the value of huge speculative loans. Thailand is just one of many recent examples of a nation brought to its knees by a currency run. Many more will follow.

Many talk glibly of the benefits of complete, rapid globalisation without any idea of the very real, human, tangible tragedies that are being created. The most vulnerable nations are making huge steps at the insistence of multinationals who refuse to invest unless they do. I have sat in meetings where the most senior members of the governments of emerging nations have been systematically bullied into scrapping regulations, in order to become more globalised. But they are being rewarded with what appears to them to be callous contempt for their own people. Of course, many of these governments have a track record of callousness themselves, despite their recent adoption of a more compassionate rhetoric. But these poorest countries cannot change as fast as the markets swing. And they have no cushion.

Borrowing to break the bank

The foreign exchange market is now an investment tool in its own right, rivalling stocks and shares. And there is plenty of cash surplus around to invest with – often in the very short term. Once currency speculators spot a vulnerable target they borrow huge amounts of that currency and sell it for another – say dollars. If the currency falls sharply, they buy back at a profit to repay the loan. Once a campaign gets under way, other investors start to panic, dumping currency and adding to the chaos. However, as George Soros himself told the Banking Committee of the House of Representatives, if everyone rushes to sell, there are no buyers, those still with positions in the market are unable to bail out and may suffer 'catastrophic losses'.

Some would argue that his foreign exchange interventions, and those of others, actually stabilise in the longer term by selling when currencies are too high and buying when they are too low. It could be argued that central banks tend to destabilise when defending a currency, because they operate in reverse, buying when the market feels a currency is already too high, and selling when it already feels too low. In the longer term it is true that free market forces operating on floating exchange rates will tend to produce greater stability – protecting a currency is usually destabilising because it risks a sudden fall or rise.

As the American economist Paul Davidson puts it: 'In today's global economy, any news event that fund managers even suspect that others will interpret as a whiff of currency weakness can quickly become a conflagration spread along the information highway.'

Governments are quickly defeated by huge flows. Every day 150,000 foreign exchange transactions take place globally, worth $2.4 trillion if both legs of each deal are counted. That's up to $642 trillion a year in a totally unregulated market open 5 days a week, 260 trading days a year, 24 hours a day. It is clear from these figures that the Bank of England, with reserves of perhaps $45 billion, could not last more than a few hours against a global dumping of sterling, let alone a country with smaller reserves. For many countries the daily trades in their own currencies are greater than their entire national reserves.

Very few nations have currency reserves deeper than the pockets of speculators and the combined panic capacity of the global market. Few can afford to raise interest rates high enough to stop large-scale speculative borrowing before they have wrecked their economies. The only other way out is to stop trying and let the currency float.

If globalisation is to proceed rapidly and harmoniously in the short to medium term, then it can only do so by a more holistic approach. Lifting all exchange controls and just hoping for the best is not enough. Wealthy and poor nations need to agree that the opening of economic boundaries will also be accompanied by support packages, to help stabilise currency fluctuations and interest rates in these countries. To some extent this is already happening after the event, with assistance for countries like Thailand and South Korea. But this route is risky and fraught with danger. Expect to see major steps taken in the interests of short to medium term global stability, with the International Monetary Fund taking the lead. Expect huge retraining initiatives, partly funded by wealthy governments, partly by the corporate sector, to assist the bottom-end losers from globalisation to become winners, gaining new jobs.

Expect growing calls for global taxes on all foreign exchange transactions so as to 'throw sand in the wheels'. Such attempts are likely to fail. The forces of globalisation are already too strong. Foreign exchange transactions are now impossible to prevent since anyone, anywhere in the world, can now trade whatever they like instantly, using the Internet and secure encryption. The black market would be vast, cheap and efficient.

Expect economic decisions to be taken increasingly out of fear of market reactions rather than out of true conviction. National economies will be increasingly controlled by the fickle and often irrational reactions and counter-reactions of the currency and stock markets. In the longer term, expect governments to take refuge in ever larger and 'safer' alliances or trading blocks, with many grouped, linked or fused economies by 2020. The aim will be to provide economic stability against speculative attack but it will be won at high cost and will not be entirely effective, as global speculators also grow with extraordinary power.

Expect large nations such as India and China to continue to observe all this activity from a bemused distance, since they already have two emerging economic areas between them serving well over two billion people. They will be hoping not to have the same market vulnerability as Malaysia or Thailand or the Philippines over the next decade, but will find themselves under pressure too.

Expect to see aggressive vocal minorities in some emerging countries, angry at what they see as fat, imperialistic money-makers who over-sell and over-buy, manipulating stocks and currency up and down for their own mega-profits. Expect these reactions to grow ever stronger as global players increase in power and weight. By 2005 there will be several institutions whose decisions are controlled by very few people, able to shake a national economy to pieces in the short term and make huge profits doing so. Expect false rumours and reports to be part of the process.

For ordinary workers in emerging countries, the globalisation of finance seems good news at first, with huge inward investment, new factories, jobs and infrastructure. Family income rises, homes are built, cars bought, university education paid for. It seems that good times will last forever and little is saved. And then comes the crash – the job goes, the loan is called in on the house leaving huge debts because of the property crash, the car is almost worthless and the children are suddenly pulled out of education. Whatever savings there were are wiped out as the value of the market falls through the floor and banks become insolvent.

And then confidence begins to return. The grossly under-priced currency once again encourages inward investment. This time foreign speculators get even better mega-deals, buying up whole factories and neighbourhoods at a fraction of their former cost. Gradually the economy recovers, jobs are created again. Life goes on. But at a terrible cost. In the big shake-outs many smaller investors, small business owners and workers lose everything, while some large (foreign) institutions make vast profits.

Overshooting will continue in currency markets

In the past one disincentive for smaller speculators was the cost of commissions, but the network society has reduced these to such low levels that multiple short-term switches are now far more common. Expect therefore that markets will also be increasingly destabilised by millions of individual decisions to buy and sell, made by people who have only a partial understanding of the underlying factors in price movements. This effect will further encourage overshooting.

For example, in Uganda many local offices of non-government organisations place overseas grants into currency accounts, and switch the money repeatedly according to their own view of what exchange rates will do. They are not finance houses and have no specialist expertise but are playing the market like everyone else – as amateurs. And large institutions in London, Tokyo and New York are playing the same game against them. After all, the only way a London house will consistently make money is from foolish decisions by others.

Every big winner means a big loser – or many small losers

For every big winner on the currency markets there are big losers. If, for example, New York teams are more skilled than those in most poorer nations and also have more clout, we can expect poorer nations to be systematically weakened by New York market activity. Of course, globalisation is here to stay and in the longer term is the most effective way to generate wealth for every nation, rich or poor, but these are some of the undercurrents which by 2010 will be causing international tension and concern.

Expect emerging nations to try – and fail – to tame the process. Expect ASEAN currencies to move together in the Far East. In the past they were all linked to the US dollar, ensuring a common exchange rate and a fix with their most powerful market. After the mid 1980s the Japanese yen rose against the dollar, making Japanese exports to ASEAN countries more expensive, as well as those to the US. The rising yen also led to massive investment by Japan in countries it had previously seen as major export markets. With investment came new technology.

The ASEAN governments were dazzled by growth and opened up their financial sectors to external agents, as well as relaxing import controls. Thailand, Malaysia and Indonesia quickly ran up large trade imbalances despite rapidly growing exports. Foreign investment was not enough to cover these current account deficits, so borrowing by industry grew. Then in 1996 the dollar grew stronger against the yen, by 30 per cent in 18 months. Japanese investment in ASEAN countries was less attractive. These were some of the factors leading to the 1997 ASEAN crisis.

The ERM debacle proved once and for all that the fixing of currencies cannot work unless it is a part of financial union. But even complete financial union will not prevent global industry pressures from controlling governments. When governments no longer control their own exchange rates, interest rates are held in common across nations, import and export restrictions and capital movement controls have been abolished, they are vulnerable indeed. Foreign capital is notoriously fickle but the only alternative is raising domestic capital, which is a major challenge in a developing nation.

Who is now running the country?

Expect more reactions like those of the Malaysian government. They recently tried to stop 'rogue speculators', thereby precipitating an even worse crisis. Expect growing frustration as these countries realise the reality, which is that they have almost completely lost control of their own economies. Expect many nations to slow down the relaxation of controls on the finance sector between now and 2010 as a direct result of market instability.

Globalisation means that large interest rate differences between nations become less sustainable, capital flows unstoppable. As the process of globalisation continues, all they will be left with is national laws and the raising and spending of taxes – but even here governments will be forced to harmonise laws, taxes and benefits. Countries out of line with what the global investment community thinks is reasonable will rapidly lose investment. Expect to see many government policy reversals as political conviction gives way to practical issues, such as whether Nissan will build a giant car factory in your country or place the project next door.

So who is now governing the country? The answer, of course, is the market, or collective decisions by some of the largest multinationals, who themselves are controlled by a few very powerful players. It's the death of democratic power.

WORKERS OF THE WORLD UNITE

Trade unionism was founded to provide protection for workers against oppression. Nation after nation was forced to legislate to provide basic rights such as holidays, sick pay, redundancy pay, maternity benefit, a maximum working week, health and safety at work, the right to challenge unfair dismissal and so on. These agreements were fixed nationally, to prevent one company undercutting another by employing workers at lower cost with no such benefits.

But that was the old world, which will be destroyed utterly by globalisation, unless we have global protection of workers' rights.

Today the same situation is emerging all over again, but on a larger scale. 'Ethical' employers (forced to be ethical by law) who treat their employees well are losing out to 'unethical' employers in the poorest nations, or will do so in the future. Globalisation means that there is a level playing field, and it's getting even flatter. The threat that jobs could go abroad lies behind job insecurity, the erosion of non-wage benefits and the catastrophic weakening of trade unions. These will all continue.

Workers in countries who enjoy a better health service, unemployment benefit, pensions and education will do so in future only if they produce lots more than those in poorer countries, and at lower cost. But how will they do so, since those in the poorer countries are likely to have all the same technology in their own factories and labour that is 80 per cent cheaper? Workers there are often paid almost nothing, have no job security, are exposed to massive health and safety risks, have no sick pay or other rights. Unless this hole is plugged, all efforts by workers to secure rights in one country will end up destroying their own jobs. It will always be cheaper in the short term not to ventilate factories properly than to ventilate to a high standard, or to give no sick pay rather than to be generous to workers who do not turn up for work.

Ironically it is often the same multinationals that employ both groups in both parts of the world, making 'ethically employed' workers redundant, replacing them with 'unethically employed' workers. It is no good assuming that workers in rich nations will always earn more because they will be more automated with higher productivity. New high tech factories employ the latest technology wherever they are built, whether in America or China. Many management, consultancy, service and support jobs will command high salaries, but that will not protect Western nations entirely from the globalisation shakeout.

Already the signs of worry are there. In September 1997 an international conference was held in Africa to address growing concerns that globalisation was a health risk to workers in African nations, with many new work-related health risks such as the exposure of women and children to new insecticides on large farms.

Third millennial trade unions must be global or die

Third millennial trade unionism can only go down one of two routes: if it stays in a national rut, fighting for national rights and for protectionism, the end will be massive job losses, and job gains in poorer nations by non-unionised, unprotected labour forces. This will be the final death of trade unionism. Every day we see the further weakening of old labour movements because they have failed to grasp the nettle of globalisation.

The only other option is for them to form a global labour movement looking to negotiate 'fair' labour rights for the entire global village. Such a formal movement would need to be capable of organising, say, global pickets of ports or global protests against particular companies. Expect some attempts at this by 2010, with little success because of the diffuse nature and great strength of globalised business companies. However, the global Internet, linking one billion people, will soon be used to mobilise instant protests of tens of thousands of people on a range of issues at more or less any point on the surface of the earth. In the meantime, informal protest movements of activists will continue to win publicity, influence and (indirectly) power.

This third millennial labour movement will be driven from the wealthiest nations, by people concerned mostly to protect their own jobs, trying to make sure that workers in emerging nations become almost as difficult and expensive to employ as they are and therefore less a threat, and also by those with wider agendas.

Of course, globalised companies hit back at the label 'unethical' when it comes to working conditions. They will argue correctly that they are providing otherwise destitute peasant farmers or street dwellers with a living wage, training and a future, building national prosperity, contributing to the balance of trade. In a competitive world, paying their workers more or giving them extra benefits could mean they all lose their jobs.

Global agreement on minimum labour conditions

Expect the beginnings of global agreements on labour conditions, health and safety and other issues by 2010, as a condition of global trade. It is already starting as a result of consumer pressure. Take for example child labour. Few household name companies today would risk 'knowingly' employing six-year-old children ten hours a day to make clothes. Yet in Bangladesh alone it is estimated that 80,000 children under 14, mostly girls, work at least 60 hours a week in garment factories.

Sadly, knee-jerk reactions and simplistic campaigns can wreck the lives of the very people they were meant to protect. In the child labour example, an international campaign led to millions of children being dumped as workers. But they and their families needed money to eat. Many children went straight onto the street as child prostitutes. The lesson is that while these issues are important, and must be addressed, they need to be tackled in the context of overall community development.

I have seen these child tragedies at first hand in India. Child labour is a very complex issue. Children with no parents, living on the street, either work or starve. It is as simple as that.

Free trade mandate driven by America

Europe and the US are the two engines behind free trade negotiations worldwide, but Europe has its own distractions and these
will continue. Therefore expect America to be in the driving seat
of far-reaching trade initiatives. Clinton played a key role in negoti-
ating final agreement on the General Agreement on Tariffs and
Trade (GATT) and in concluding the North America Free Trade
Agreement (NAFTA). Watch out for more from the World Trade
Organisation (WTO), Free Trade Area of the Americas (FTAA),
and the Asia Pacific Economic Co-operation forum (APEC) whose
18 members comprise half the world economy. APEC has agreed
to achieve 'free and open trade and investment' by 2010 for industrialised members and by 2020 for the rest.

Expect organised labour movements worldwide to campaign
increasingly against free trade and in favour of hostile trade blocs.
Expect such voices to grow powerful in the US during the Bush
administration. The US has not had a protectionist president for a
century (apart from President Reagan's spate of import quotas) but
could yet find itself with one in future. If the US becomes protectionist (possible but unlikely) then expect a severe defensive reaction
from more than 100 other nations. Big trade agreements will continue without America if necessary, but with the serious risk of
becoming derailed by some other anti-globalisation backlash.

Old economics is dead

Old economic theory was built for a world which no longer exists.
It stated that, for example, 'high unemployment causes wages to
fall', and 'low unemployment causes wages to rise'. But in 2000
unemployment in the US was less than 5 per cent without inflationary pressures. Why?

The old boom and bust economics is nation-based and semi-
redundant in a third millennial world, where forces are infinitely
more complex. We will see big booms and big busts and disruptive
business cycles, but their causes and time-scales will not be as easy
to manage or predict as before.

Two factors have combined in the new world order: firstly, globalisation means global competition which is now restricting the

ability of workers in any one country to demand higher wages. In the global village they know that another, less greedy country will win the contracts. Secondly, computers and other technologies are triggering a massive increase in productivity along with rapid downsizing in manufacturing, and deflation of costs.

FACTORS AFFECTING REGIONAL GROWTH

Geography matters – or does it?

Recent studies by the Harvard Institute for International Development have found that global growth from 1965 to 1990 depended on four factors:

- initial conditions
- government policy
- geography
- demographic change

In the past, different policies in different regions produced huge differences. In the future, while policies will tend to converge, many developing countries will be left far behind, partly because of their geography. It will always be true that sea trade is cheaper than long-distance land (or air) trade, as Adam Smith first recognised. It is no accident that 150 million people in China are on the move to more rapidly growing and dynamic coastal provinces.

Throughout history coastal states have tended to develop freer market policies than their landlocked neighbours. Mountainous states have neglected market trade due to isolation. Temperate areas have had larger populations, able to feed industrialisation, and tropical nations have been burdened by far higher death rates from disease and low agricultural productivity. Torrential rains in tropical areas erode the land and leach nutrients where forests have been cleared.

Despite the transport and communications revolutions, landlocked countries will tend to grow more slowly, as will those in tropical areas, those with corrupt or inefficient governments and those with the highest population growth. All these factors consume resources that could be spent on production.

Population grows as medical care improves. Then birth rates fall as households adjust to longer life expectancy and lower infant mortality. East Asia is into the second phase while Africa is still in the first. Thus Africa has a huge bulge of dependent children (up to half the population in some areas is under 15 years old) with no bulge in workers, a problem made far worse by the selective destruction of young adults by AIDS.

Two of the most effective ways to help Africa would be to tackle AIDS with far better aid programmes, including treatment of other sexually transmitted diseases, and to tackle malaria. Malaria kills 1 million a year and affects between 200 and 400 million others. Ninety per cent of the deaths are in sub-Saharan Africa. Despite this, total global spending on public health malaria programmes is a mere $60 million a year and pharmaceutical companies have invested comparatively little. A simple, low cost malaria vaccine could revolutionise large parts of Africa. Expect an effective vaccine to be developed by 2007.

These huge contrasts in wealth and health in the global village will make large, uncontrollable population drifts more likely, creating all kinds of tensions in the future.

Companies switch many times from country to country

Many US toy manufacturers have moved their factories from Japan to Taiwan to Singapore to Thailand, chasing lower labour costs – finally ending up in China. In contrast, Lego still produces most of its toys in Denmark, Germany, Switzerland and the US because it needs high quality injection moulding and mould design.

Some factories are in other countries where labour costs are low. Others are source factories, with skilled local managers able to source materials and redesign processes. Others are server factories supplying specific markets so as to overcome tariff barriers, taxes, or the logistics costs of foreign exchange risks. Then there are lead factories creating new processes, products and technologies for the whole corporation globally.

Sunlight is the greatest barrier to the global village

Globalisation means long-distance travel for executives, because technology has not yet caught up with the movement of capital and other components of international trade – or rather the technology is there but the social skills are not. Most pre-millennialists just can't cope with electronic meetings, but their jobs will depend on them.

The greatest barrier to the global village is sunlight. In an ideal globalised world every inhabitant would be on the same time clock. And that is the crux of a growing problem. The third millennium will see a whole new pattern of working. Days of 9 am to 5 pm, 8 am to 6 pm or even 7 am to 9 pm will be replaced by a different rhythm, dictated by business efficiency and what globalised customers and corporations want.

Consider the example of a private banking client who in an ideal world would like to call her banker from wherever she has just flown today, without having to look at her watch and do a mental calculation across a time zone. Ideally she wants to be able to talk anytime. She can. Omnipresent technology such as the video dataphone, coupled with tomorrow's satellite technology, will allow her banker to be constantly available by phone, fax, video link. But what happens to his life?

The answer is that work patterns change. Instead of having, say, five very important clients that he services during traditional hours, he now is available at all hours, but less intensely. As compensation for offering such a premium service, he can afford to have only four clients and still have the same income. Most of the work is still done during normal hours, but out-of-hours unscheduled work and pre-booked meetings are compensated for by shorter hours overall, and time off during the day.

So he might be playing golf all day Monday. His clients are educated to know that this service means that when they phone he could be anywhere, doing anything. But with a dataphone almost all the information he ever needs for dealing with a call will be available in a minute or two, if not in seconds.

Many people already live like this, myself included. For a start, anyone involved seriously in the media as a commentator on global events has to. That is because news stories can break at any time, day or night, 365 days a year, and when they do, media researchers need instant access to expert comment and advice. Television and radio are particularly demanding.

Take the example of the cloning of Dolly the sheep, one of many major news stories I have predicted over the years. The real story was not about sheep but the possibility of cloning humans. Two TV crews were at my home at around 11 o'clock on a Saturday night, within 40 minutes of their first telephone call, just as the news hit Reuters and the Press Association. But then that's a part of the world I live in and so I budget time to allow for it. That means running the diary with fewer pre-booked meetings, with flexibility and availability.

Living 'on call' is nothing new. Doctors and many people in other professions have been used to it for decades, as have all those who organise themselves so that (after the proper screening of calls) they are very rarely if ever uncontactable.

For holidays and special family times phones can be carried by others, calls diverted and contingency plans laid. Cover can be arranged – and should be. We are not talking about communications slavery with technology, but about using technology to liberate us from the pre-millennialist expectation that we should all enjoy full-time leisure interspersed with very intensive full-time work.

Some say that this new pattern of global time-keeping is unhealthy or unnatural. The reality is that it is far more in tune with the old hunter-gatherer pattern of life, and of course identical to normal patterns of life for mothers or fathers at home alone with several small children.

Life in those circumstances is not a set of neat on-off buttons. But it does involve a different mindset from that of the conventional office worker. It also has implications for where we work: when executives need to take or make calls very early or late in the day, or over weekends, the trend to home working is accelerated.

Without an attitude change the result will be burnout and the destruction of families. We are already seeing workaholic techno-freaks with portable phones and computers sitting on beaches on holiday. Office addicts who cannot relax. Techno-junkies driven by insecurity about what might be happening back at the office while they eat at a local taverna.

But without a massive rethink about hours and daylight, global village life is never going to work without East/West discrimination in favour of North/South alliances. It is already happening. It is far easier for a company based in South Africa to work in close partnership with Britain than with Hong Kong, from the point of view of working hours. The problem of daylight incompatibility is made even worse by cross-cultural differences. A company in San Francisco trading with Dubai finds not only a disruptive time difference, but also that Dubai works from only 7 am till 1 pm and does not work on Fridays – but works a normal day on Sundays.

Globalisation will be patchy at first

Expect big differences and inconsistencies over the next two decades in the degree of globalisation in industry, as in the last decade. For example, while cross-border food and drink deals worldwide rose to $6.82 billion in the first half of 1997, from $5.34 billion in 1996, UK food and drink manufacturers actually reduced their foreign investment from $176 million to $138 million in the same period. Foreign buyers also melted away slightly, despite Coca Cola's takeover of a joint Schweppes bottling venture for over $1 billion. Expect many big mergers in the food industry, forming globalised giants with combined turnovers in excess of $15 billion who are shaping up to dominate the global market. Expect more hostile takeovers like that of Docks de France by Auchan in France or the moves of Carrefour towards Cora, or Promodes with Casino.

'Future Branding' – universal and tribal

'Future branding' is the reshaping of a brand not only for today's needs but for tomorrow's globalisation. For a while British Airways abandoned its British-style logo for a new brand image as a global carrier. The tail emblems were based on designs from different nations. Their chairman predicted that the word 'British' might even be dropped in the future. So BA tried a global image to go with a global product. But many fly BA precisely because it is British (tribalism). British Airways has a strong, staid, conservative, ultra-safe character in many people's minds. Of course you could launch a new brand altogether: 'Global Airways for world-class travel' – but it would not be 'BA – the world's favourite airline'.

The British Tourist Authority (BTA) tried to do the same thing, announcing it would drop the Union Jack symbol in marketing Britain – while other nations such as the Irish were reverting to theirs in tourist literature. The BTA also soon changed their minds.

Expect a wholesale rebranding of national airlines throughout the world, together with consolidation. There is no room in the market for every nation to have an unsubsidised privately owned 'national' carrier. Country images inspire loyalty, hostility and a range of other emotions and national images change. Swissair and El Al are two other airlines with strong national images, sporting their country's flag or colours. These airlines will be thinking again about their national image, as will airlines based in other transforming nations such as South Africa, Saudi Arabia, Spain, India, Malaysia and Morocco. As airlines become multinational corporations they will need to think and act globally. Many national airlines are struggling to develop and embrace a new global culture in competition with low-cost budget airlines with higher efficiency.

VIRTUAL COMPANIES

Globalisation means subcontracts and partnerships at every level. Expect more virtual companies: companies that have far fewer employees than you might expect from their global operation and high turnover.

According to *Vision*, a study by the Economist Intelligence Unit and Andersen, 79 per cent of senior executives at multinationals believe that large business structures will change by 2010. The future will be virtual companies – 13 per cent of all business was already 'virtual' by 2000. Working practices 'on the ground' may already be more virtual than many board members realise:

- employees in many different locations
- rapidly changing structure
- many functions carried out by partners.

While only 3 per cent said their companies were fully fledged virtual companies in 1997, 40 per cent of senior executives said their companies would be virtual by 2010. Ninety per cent of executives know they will need more sophisticated communication skills to cope and 80 per cent say their companies lack the necessary relationship-building skills to make virtuality work.

ILAN Systems Inc. is a business founded in 1993 with more than 50 employees and sales approaching $6 million. The president, Tom Reynolds, has his headquarters in a couple of cramped rooms in his 1,500-square-foot house in South Pasadena. His partner and other administrators also work from their own homes. They keep in touch by phone, e-mail and their own computer network. It's flexible, responsive and helps companies react fast. ILAN pays salaries, gives holidays and provides health insurance.

Other companies have moved a stage further: no permanent staff. They hire freelance workers, typically home-based, on a project-by-project basis, in the same way that Hollywood puts teams together to make movies. Roles can be reversed when several groups of people have several companies between them. In slack times for one company, a director may find himself contracted to a new project run by a company belonging to one of his or her own freelancers.

Around 50 million Americans are now working from home, part-time or full-time, using e-mail to work with others. The spread of virtual companies will accelerate this.

However, virtual working already impacts upon many families and marriages, with people working long hours and a blurring of the gap between work and home. There is often no cover either for sickness, training or other traditional employee benefits. The up-side is that productivity can be sky-high with self-motivating employees. There is no paper shuffling or hanging around the door for a chat. All you can do is work.

Virtual companies avoid high risk start-up costs, such as were seen recently with biotech initiatives. When biotech started, companies built scientific teams, rented big laboratories – and then caught colds from big overheads when facing problems getting their products to market. In contrast, the biotech virtual CEO works in an office by himself with everything contracted out. The company is generally in a university and pays no rent; drug discovery is carried out at additional universities under contract.

Metacrine Sciences Inc. (MSI) is another example, run by Jeffrey White as a virtual company since March 1996 from a simple office in a New Jersey university. All the research is carried out by academic departments in return for payments or equity share. But how do you convince a big corporate backer you are real when they want to come and visit?

BP recently introduced Virtual Teamwork to save travel and manpower costs. On one North Sea oil platform construction, BP saved $4.5 million (£2.8 million) in one year for an investment of $500,000, by communicating far more widely than before. BP staff using technology are now far more likely to consult both inside and outside the organisation.

Knowledge management is the key to virtual working

Knowledge management will make all the difference between survival and death in the future. New specialist posts are springing up everywhere: Chief Learning Officer at General Motors, Group Director of Organisational Learning and Development at UBS, Chief Knowledge Officers at Skandia in Sweden and Chapparal Steel in the US. It is all part of the same trend towards marshalling information resources and making the right decisions quickly by using multiple channels and partners.

Many of the largest manufacturers will push towards decentralisation and subcontracting. Mercedes-Benz gave 40 per cent of the design work on a new Swatch car to outside consultants and 75 per cent of the vehicle development to component module manufacturers. As part of the process, niche globalisation will increase: there are now only three large suppliers of car seats in the whole world.

Customers count more

Expect a further huge shift to customer/client contact in all areas, whether health, law, retailing, financial services or any other area. Everyone has customers: external or internal. Expect more selling of products and services by one department to another in the same corporation, and then further outsourcing as the real costs of internal provision are exposed.

Adding shareholder value – but what is real value?

As shareholding becomes more globalised, pressures from share-holders will change. Adding shareholder value will continue to be important, but this emphasis will encourage short-term return on capital rather than long-term corporate strength. Short-termism will be made worse by remuneration packages which continue to depend on last year's company figures, rather than on 360-degree assessments, where each worker is asked not only to assess the performance of juniors, but also of peers and superiors. Expect a shift from such year-on-year blindness as investors increasingly recognise longer-term assets, such as work in progress.

In future a company may only be as good as its shareholders, with more board decisions governed by what shareholder reaction is expected to be. Expect large institutional investors, such as pension funds, to become increasingly involved in major corporate decisions, effectively hiring and firing the most senior executives and exercis-ing a veto on new policy. But these institutions will not be well placed to judge what is best for the longer-term future of the company.

Already institutional trades account for almost 90 per cent of the volume and value of trades on the New York Stock Exchange (pen-sion funds, investment companies, foundations, mutual trusts and banks). Institutions will continue to own the vast majority of US stock beyond 2010.

Company valuations will continue to develop as a specialist blend of art and science, with an increasing number of 'soft' variables such as the intellectual capital of the current workforce, and its future likely worth in the light of expected staff turnover and possible 'brain poaching' by rivals. The extra value derived from such things as completed internal reorganisations and the re-skilling of staff will increasingly be recognised and debated.

Virtual corporations

A virtual corporation is a collection of companies, some perhaps virtual themselves, organised to behave as if it were a larger, multi-faceted organisation. Expect more of them. Headquarters will no longer indicate size or profitability, only how hierarchical the com-

pany structure is, and how much money the shareholders are willing
to waste on a prestigious monument to past greatness.

Virtual corporations can be seen only on paper, video or the
web. The nearest you will come to visiting one will be at a large
corporate gathering, attended by key people from across the world,
representing every aspect of the virtual structure. At first many
people will refuse to take virtual organisations seriously, until they
find their profits and jobs disappearing. If virtuality means almost
the same output for less cost, and a faster response time to major
changes, then what is the point of not being as virtual as possible?
Expect all kinds of high tech experiments, for example CEO offices
lined with a dozen screens so that at any time people, data and
other images are on display, including perhaps the view in the virtual
corridor outside.

SUPER-CORPORATIONS – RIVALS TO STATE POWER

Expect the opposite trend as big companies merge to become a
colossus of global expertise and economic power unrivalled on the
face of the earth, more powerful than most governments.

The process of globalisation has hardly begun. In the 20 years
following free capital flows and global networking we will see new
super-corporations, each larger in economic weight than many
sovereign states. These complex institutions will have the power to
dictate terms to governments and set the agenda for commerce,
and will attempt to create global monopolies. They will be beyond
state control, so globalised that their power base shifts geographically whenever it suits, and will provoke protests.

What happens when a single investor has the capacity to
wreck an economy or resurrect it? Those days are already near.
Calls for super-state control will grow stronger as global monopolies grip more tightly. This will become a key issue for governments, who will find themselves outmanoeuvred by rapidly
changing corporate decisions, and also for voters who will be dis-
empowered.

A government might take a year to change or reverse a particular policy, while a corporation may take a major decision as a response within a day. Twenty large super-corporations with unimaginable combined economic power will make many of tomorrow's governments look like dinosaurs, struggling to comprehend the world around them, to react fast enough. This will be particularly the case in emerging nations. And while remuneration packages for CEOs in industrialised nations remain so much higher than for prime ministers or presidents, many governments will also find themselves crippled not only by cumbersome decision-making machinery but also by serious lack of brain-power. The brightest and most talented leaders of tomorrow will tend to go to where the real power is. They will choose to run corporations rather than pretend to run countries.

While many super-corporations will dominate specific fields, such as satellite broadcasting or computer hardware/software production, they will also show a remarkable ability to jump into unrelated fields, and to dump old businesses.

Mergers and acquisitions – and liquidations
Expect a new wave of mergers, demergers and acquisitions, with the chaos increasing in a huge global realignment. Expect many 'buy and dismantle' operations, where the break-up price is greater than the complete sale value. Many companies that might just have turned around will be ruthlessly liquidated, producing a fast return on capital.

Big shakeouts
Globalisation means big swallows small. It means huge chains gobbling up thousands of small independent outlets. It means huge discounts because of greater buying power, lower prices on some goods and inflated prices on others because of virtual monopolies in some areas.

Microsoft is a good example of a corporation with money to spare. At one point the Seattle-based software house was sitting on $9 billion in cash, growing by $18 million every day. A major challenge for them is what on earth to do with it all to ensure a proper return. Watch out for a string of new acquisitions and more anti-monopoly challenges.

Globalisation and consolidation

Industry after industry is facing big shakeouts. Once-stable markets like banking are shedding weaker players and merging into new big groups. Software houses are seeing faster and more intense shakeouts while their sales grow. The personal computer industry has already shrunk from 832 to 435 companies in the late 1980s and may soon shake down to a mere five to ten long-term winners. Likewise the dot com shakeout will be savage before the next leap to new technologies and networks.

Seismic shifts have been hitting mature industries. Examples include drug companies before governments began to insist on generic prescribing, and before global pressures forced them to concede patent rights to poor nations, banking before deregulation and the arms industry before the collapse of communism.

Deregulation of banking

In 1985 the US had more than 14,500 commercial banks, mostly operating by law in only one state. By 1995, limited interstate banking allowed this to drop to 10,000. This will probably fall to less than 2,000 as further deregulation allows national banking. In most countries of the world expect the number of conventional banks to be reduced dramatically, especially in heavily 'over-banked' nations such as Germany and Switzerland. Banks in countries like the US, UK and Switzerland will go on making deep cuts in their labour forces, at a time of great overall profitability.

There will be a huge spate of mergers and acquisitions and unhappy shareholders will face falling returns on equity in the future. The retail banking sector will be barely recognisable by 2010, and virtually invisible on the High Street apart from the ubiquitous cash machines – yes, cash will live on.

Cash will survive well into the second decade of the third millennium but goods bought with cash will cost more because of the growing costs of handling cash. The exception will be where cash is used to avoid tax. All institutions converting electronic money to and from physical cash will charge ever-increasing amounts for doing so, encouraging cash to flow around without ever being banked. Cash will continue to be the main payment method for black economy transactions, which will still be numerous but of low individual value. Because of this, cash transactions will increasingly be associated with deliberate tax avoidance.

Don't believe your customers

Market research, with its intense interest in the consumer, was the big thrust of the eighties and nineties. But as we have seen (pages 7–8), the trouble is that consumers only know what they know – which may be a lot less than you.

The use of portable computers on aircraft is a case in point. Big airlines did expensive surveys which all showed the same thing: only a minority wanted all-singing, all-dancing technology in-flight to make them work even harder. For many their flight was a welcome break away from the fax and phone. And as for using portable PCs, only a few people had them, and even fewer could imagine themselves spending an entire flight several hours long at the keyboard. So, after much hesitation, airlines decided to experiment with phones (and faxes in some cases) which were introduced gradually, on business first class. But they were wrong.

Consumers changed their minds. Many executives had no idea just how vital their ultra-powerful portable PCs would be for writing speeches, or preparing last-minute adjustments to an important presentation. As executives fly more, they can no longer afford in-flight leisure time.

Phones remained a minority interest – not least because some airlines made another huge mistake when they installed phones without a standard socket for attaching a PC, or that had connection methods which were difficult to use. Executives wanted to glance quickly at their e-mail, download a document on the way to a meeting, send e-mails around the world, not to use a voice line.

But the biggest mistake of all was to commission brand new jets with no power sockets in the seats. Those that have done so now enjoy a competitive advantage over the others, especially since spare computer batteries are bulky, heavy, only last three hours, are a nuisance to recharge and cost up to $300 each.

Moral: listen to your customers but don't *believe* them when it comes to the future.

Keeping in Business

Early warning signs of a big shakeout include:
- Belief in a company that it can't happen – it probably will
- Entry rates of new competitors
- Excess capacity in industry
- Pressure on margins as prices drop
- Scenario planning based on different forecasts
- Identifying which competitors are likely to survive best

GLOBALISATION AND NEW TECHNOLOGY

The end of mass products?

Factory production has been based on economies of scale – big factories, high volume, low margin. But new technology allows smaller players to enter, make niche products at low cost and move on before the old dinosaurs have even had time to react. Thus small new players will continue to enter markets successfully, despite domination by super-corporations.

Mass customisation will be a key feature of the next millennium. In a sense it's nothing new. Back in the 1970s you could order a Ford estate car with any one of hundreds of variations like quality of trim, paint colour, type of radio. Every car on that mass-production conveyor belt was built to order. I walked up and down the four-mile line and watched it done. The final car rolled off the belts with a delivery name and address neatly packed in a polythene sleeve attached to the windscreen.

The most obvious new area for mass customisation will be interactive TV, where viewers will find themselves targeted with adverts specially selected for them. Digital TV will copy what is already happening on the Internet. With so many millions of pages to browse around, it is easy to place ads on just the right pages to fit with your products.

Disintermediation

Globalisation and technology such as the Internet will combine to wipe out middlemen, placing customers in direct contact with sellers. Examples already include shopping for air fares: the Net can do a global search and give you the best deal in the world in seconds, and there will be two prices for everything, the human deal and the Net deal – and the Net is always cheaper.

Another example is estate agents. In the early 2000s hundreds of property companies appeared on the Net, selling houses. Some were entirely Net-based while others were traditional estate agents, offering Net services as 'added value'. Estate agents are always going to be with us but their approach is being altered by networking. Expect travel agents, car retailers and a host of other retail industries to be hit by a big shakeout. In future the most contact many companies will have with their consumers will be via the customer delivery van.

Globalisation will proceed far more rapidly than virtualisation

When e-mail arrived, people still printed out paper and filed it. Paper systems continue because people still send letters and faxes. Scanning into virtual filing cabinets is still a rarity, because it is still more expensive and less reliable. Most businesses will continue to use paper-based filing systems well into the second decade of the third millennium. It's the same with virtual meetings. You may be happy to operate that way and insist that all new employees do so, but your most important clients may not be.

Technological discontinuity

Paradigm shifts hit hard. Look for example at office equipment suppliers such as Addressograph-Multigraph Corporation, a huge company which died at the feet of the personal computer revolution.

Even NCR struggled to adapt in time. The trouble is that by the time corporations realise, putting right the mistake is near-impossible, or too expensive.

Travel industry revolution

The travel industry is about to be hit by enormous pressures from new technology. An airflight is an airflight, so long as I get the same ticket with the same carrier in my hand. On the Net a travel agent in Peru is on the same playing field as an agent at the end of my street.

The year is 2005: I log on to my PC and tell my intelligent agent to book me the cheapest ticket with one of my favourite airlines from London to Paris. A few moments later it proposes an order, which I click to confirm. I swipe my smart card through a slot in the PC. (It also holds my medical data, bank data and everything else. I have no other cards now.) The card has been loaded with my ticket details and seat allocation and has also provided payment.

When I arrive at the airport, they swipe the card, which also acts as passport and visa holder. A holographic image of my face appears on the screen and is checked by the only human I meet on security side. At a different airport I stare at a camera which analyses the pattern of blood vessels at the back of my left eye so the robot knows it's me. On another occasion I have no time to book and literally run onto a flight with my hand luggage. As I enter the aircraft door, the smart-card chips in my wristwatch, or under my skin, authorise an automatic debit to my account to pay for the air fare. Sure, many people will still do things the slow and expensive way, but life will change far faster than you think, especially when all this can be organised while watching cable TV, from any public phone or from any phone in your own home.

Sea freight revolution

Globalisation stands or falls on low cost international freight, which usually means container ships. Freight owners will benefit from the network revolution. Shipping is dominated by the manifest: paper documents to prove who owns what and show where it is headed. Ports are bedevilled with paper and paper is inefficient. Just working out the loading and unloading order for hundreds of containers in a large ship used to take a couple of days by hand.

Now a ship can be loaded to order, with complete precision. Bar codes on every container, and laser readers as well as transponders mean that even in the largest port the location of every container can be identified on a screen to within a few metres. A client anxious about her freight can punch in the details on the Net and see where the goods are – in the port, on the high seas – and receive accurate information about the timing of the next leg of the journey.

Four Key Technologies

Freight is being taken over by four new technologies.

- Portable computing – pen-based and touch sensitive. Tick the boxes and away you go
- Radio frequency equipment – instant wireless data transfer between everyone everywhere. The gantry operator 150 feet up in his cab has instant electronic data from trucks and ships below, as well as a constantly updated loading plan
- Bar codes and overhead scanners – tell everyone what cargo is where. Internet/intranet will allow instant billing, orders and information on freight location, loading and unloading so lorries arrive in the right order at the port
- Intelligent tags/clips – attached to containers or their contents allowing precise monitoring at speed. They transmit information to sensors as they pass.

Freight theft

These new technologies cut fraud and reduce theft ($10 billion in cargo goes missing every year in the US). Theft is such a problem for companies like Digital that they have stopped labelling their own containers. In a short period they had 129 container thefts in 13 countries.

Compare the costs below:

Gold $350 per ounce
Cocaine $650 per ounce
Pentium chips $800 per ounce

So which item is targeted first by the thieves?

Air freight set to grow

Although sea freight is cheap, in the case of a high value, low weight, time-critical product it is often not fast enough to enable a distant producer to compete with local companies. Hence one requirement of globalisation is rapid cargo delivery. Speed brings savings in storage and enables just-in-time solutions. High value goods such as computers, software, medical equipment and other time-sensitive goods such as music CDs are increasingly being carried by air. Air cargo doubled between 1987 and 1996, and accounts for a third of global freight value. Expect a further doubling over the next five to six years.

GLOBALISATION IS CHANGING COMPANY STRUCTURES

The old-style company began as a kind of a pyramid:

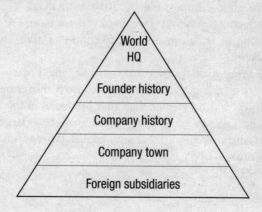

And the subsidiaries really were foreign, always subservient, lacking in power, controlled from the top down. This kind of structure is increasingly irrelevant in a globalised world where location becomes less meaningful. The next structure is the Universal Global Web:

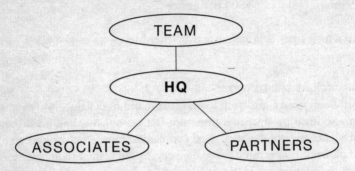

Globalisation forces corporations with strong tribal (national) identity to ask 'Who is us?'

IBM is well known as an American company, yet 40 per cent of their staff are now outside the US. It could be the majority before long. IBM Japan alone employs 18,000 and exports goods worth $6 billion a year, much of it back to the US. Dutch company boards now have triple the number of non-Dutch directors compared with the late 1980s (now 12 per cent). What happens in another decade? Many companies, wherever they have their HQs in the world, will have boards where the majority are from other nations. Indeed it will become almost impossible by 2010 to manage a globalised company effectively without globalised representation at the top, partly because of tribalism and a perception of staff and public image problems.

By 2010 many strongly national companies like France Telecom will have very diverse global operations, where turnover outside the 'home country' will be greater than that produced in the domestic market. Expect a spate of company name changes as large corporations refit their images to become global citizens rather than national entities.

Cost of capital converging

Financial controls have crumbled so that the cost of capital is converging between countries. Factories can be located anywhere, linked by satellite.

A cotton trader has a business in Pakistan buying cotton and making clothes for export to the US. Along come import quotas designed to protect US jobs from the global market. He can't export any more clothes from Pakistan but Zimbabwe has plenty of unused quota which he can buy off other traders. Within twelve months the trader has set up a new factory in Zimbabwe. The cotton is still harvested and processed in Pakistan but shipped from Pakistan to the factories in Zimbabwe. If export restrictions are lifted the Zimbabwe factory will close almost overnight.

Workforce left behind

The trouble is that such factory shifts as these leave people behind – except for a few specialists who move as well. People are less mobile than capital, technology, information and raw materials. Skills in a workforce are a vital national asset. Imagine trying to create a new Silicon Valley in France, attracting hundreds – even thousands – of key technology personnel there from the US. It would be almost impossible. If you want skills, you move your site to where those skills are. Hence Bill Gates decides to create a huge new £50 million research facility in Cambridge, England, run by British professor Roger Needham, recognising that no amount of cash will woo the best British brains in sufficient numbers to his own sites in the US.

A recent study suggests a direct correlation between big increases in home ownership and rises in unemployment. Compare for example Finland, Ireland and Spain with 78 per cent ownership to Switzerland with less than 30 per cent. Home ownership makes moving more difficult and expensive. In France, Italy, Spain and Belgium the cost of selling one house and buying another (estate agent fees, legal costs and stamp duty) amounts to more than 15 percent of the price of a home. In other words, a 15 per cent gain over three years could be almost wiped out if someone moves three times a decade. Home ownership is less of an issue in the US, where buying and selling is easier.

Home ownership will continue to be popular, but owner-occupation may decline. Expect many home owners to become absentee landlords, renting out their own homes as investments while they are working in other cities or countries. As we will see, hypermobility is very costly in personal terms, leading directly to the break-up of marriages, the scattering of children into various countries because of fixed educational commitments, and the loss of other long-term relationships.

Executives need more than money to move

But if people get left behind as industries come and go with remarkable speed, shifting themselves around the globe, then retraining is going to be an even more important national priority. Companies may not be able to guarantee jobs in the future, but at least they can invest in their staff so that when the company moves on, it leaves behind employees who have given of their best right up to the split, knowing the company was also doing the best for them by investing in their future.

Many middle-ranking executives need a lot more than money to persuade them to move. It is true that a starving man or woman will travel a long way to win daily bread for a family. But in many Western nations where most people have a reasonable standard of living other factors become more important. Relationships and family keep many rooted. Parents with teenagers at a critical stage in their education are often extremely reluctant to go. As are those on their second or third marriages, having learned painful lessons from the past about neglecting home life, perhaps now with a 'new' set of very young children they are determined not to neglect as they did the set who are now grown up. Increasing numbers of older executives in very senior positions are also bound to a locality by responsibility for elderly parents, at a time when their children have finally left home.

Lower down the social scale there are armies of mobile workers in countries like India who are used to spending eleven months a year away from home, earning money to support a wife and family. Many go farther afield, for example working in Dubai as taxi drivers, returning only once every two to three years to see loved ones. But this is globalisation.

VAST AND INVINCIBLE

Small may be beautiful but big means powerful. Who is going to regulate the new super-corporations? Monopolies and oligopolies have hardly formed yet on a global level – globalisation is too young – but they will. As we have seen, vast multi-faceted international conglomerates will have more muscle than the largest governments. Expect growing calls for their regulation. But where is the global control going to come from? Expect this to be a key issue by 2005, with attempts at regulation by 2010 – which will only be partly successful.

CHALLENGES TO MANAGEMENT

Globalisation as an issue
◆ Has your company recently addressed the new challenges posed by globalisation?
◆ Are all senior staff up to speed with the extraordinary changes in the global economy over the last three years, and the likely impact beyond 2005?
◆ How globalised is your head office in mentality?
◆ How would you know?

Currency instability
◆ How vulnerable are you to increasing political risk and currency instability, especially in emerging markets?
◆ What 'insurance' have you to cover this risk?
◆ Are you prepared for very hostile attitudes to foreign investors who rapidly remove capital?
◆ How ready are you for government policy reversals at short notice, precipitated by fear of the markets?

Global pressure from labour force
◆ How prepared are you for labour movements to join forces in several countries against you?

Working hours in a global village

◆ What is your expectation of staff that are required to interact with colleagues very early in the morning and very late at night, because of time differences?

◆ How should your office hours culture adapt in the longer term and what are you doing to encourage this?

◆ What is your policy on home working and mobile working (e.g. dataphones) and how does it need to change?

Rebranding

◆ Do your products need rebranding for a global market?

◆ Do your 'global brands' need better management as a portfolio of niche brands?

Virtual companies

◆ How easy would it be for a virtual company to set up in competition with you and take some of your business?

◆ Are you as virtual as you should be, given the possibilities created by workforces and new technology?

◆ How could you be more virtual and more profitable or efficient?

Shareholder value

◆ Is your company being valued correctly, e.g. by the use of new measures such as intellectual capital?

◆ How well are you managing shareholder expectations regarding the need for longer-term views of success than last year's balance sheet?

◆ Where are the threats and opportunities in mergers, acquisitions and disposals?

◆ How vulnerable are you?

◆ How well can you survive in a huge global race on your own?

Disintermediation

◆ Is your company fully exploiting new technologies to cut out 'middle merchants', with direct buy and sell?

◆ Are you vulnerable to being 'cut out of the deal' in the future?

Switching countries

◆ Is your location policy due for review?

◆ How certain are you that company activities are all located in the right country?

◆ Do you have rapid exit strategies from countries which may become unstable?

Market research

◆ Are you using psychographics and other new technologies to predict what customers will really want tomorrow when surveys tell you only what they can see from today?

◆ Do you have good access to accurate global trend forecasting, taking into account discontinuities?

Company structure

◆ Is your company structure changing as fast as the world you operate in?

◆ What is the future of your HQ in a world where partners, associates and competitors are networking continuously?

◆ Does your current structure give you a competitive advantage – or is it a fudge?

Company identity

◆ How important is the current image of the company, and is it the right image for the third millennium?

◆ Is national identity an advantage or disadvantage?

PERSONAL CHALLENGES

Personal globalisation

◆ How globalised are you in your own thinking?

◆ For example, do you read a journal like the *Economist*?

◆ Do you watch a global news channel such as CNN?

Multi-skilling

◆ What would you do tomorrow if a big merger or acquisition meant that your company and role was likely to disappear?

◆ What kind of insurance policy have you taken out against unemployment – for example, do you have a second or third skill which you could use to get a job?

◆ What can you do now to keep yourself in the mainstream job market?

Multi-language and multi-English

◆ How many languages do you speak, and do you need to improve your language skills? If English is your mother tongue you may not understand the need. English may be the dominant global language but we are talking about building relationships, communication and confidence raising.

◆ Is your English world-class standard or could it hold you back?

◆ Do you understand the difference between your own version of English and international English? The words used, the grammar and accent are entirely different from British or American or Caribbean or Indian English.

◆ Are you speaking the right version of English at international forums?

◆ Do you speak clearly or do you mumble?

Moving to a virtual day

◆ What steps have you taken to move away from a traditional working day with a beginning the middle and end – towards working across several time zones? That means taking leisure time in the middle of a traditional day in the office, and being on call for up to 18 hours a day. It also means taking whole days off – apart from being on-call – during a 'normal' working week.

◆ How prepared are you for virtual working?

◆ If your company had to choose the most virtualised team members for their next initiative, would they choose you?

Mobile home

◆ How mobile do you want to be in the next five years?

◆ Is that compatible with the personal commitments and preferences of other members of your household?

◆ Have you talked those issues through enough to have a joint plan?

◆ Have you considered developing alternative career paths so that you can have the option to stay where you are rather than relocate if your job changes, or to be able to relocate because your partner is relocating?

Greater knowledge means more power

◆ Do you know how to use your company intranet?

◆ Do you have a thorough understanding of how knowledge is managed inside your company?

◆ How valuable is your own contribution to this growing knowledge base?

◆ If staff were rewarded according to how many times other people accessed their own unique pages of information, would you rank as a net provider or taker of data?

And finally . . .

◆ How vulnerable is your own portfolio of investments to a major currency crisis?

CHAPTER 5

Radical

Reacting against twentieth-century values

The future will have a strong radical element to it, as traditional political movements shrivel and die.

GOVERNMENTS – LOSS OF POWER

All over the world the political scenery is changing and old forces are dying. What will replace them all? Governments are losing power. As we have seen, the global economy rules entire nations. Trading areas are now regional. Currency is decided by the markets. In addition, state services have been privatised and there is a trend towards local government.

In the case of Britain, most laws are now made by Brussels and national decisions are overruled by places such as the European Court of Human Rights in Strasbourg. Loss of power upwards is balanced by loss of power downwards with a new Scottish parliament, a Welsh assembly, and the devolution of power to Northern Ireland. Westminster will be left as the mini-parliament of Little England – minus London and other large cities, who increasingly want local powers of their own.

At the same time government runs less and less of the economy. Gone are ministerial responsibilities for water, electricity, coal, gas, British Airways, telephones and much of public transport. This trend to privatisation is now global and unstoppable, even taking deep root in communist China. There will be major challenges for corporations to run these privatised industries. We will see growing concern as nations lose control of their own utilities through international ownership, with, say, a gas company in America buying a huge stake of an electricity provider in France.

Non-profit organisations set to grow

The welfare state will be privatised extensively, to non-profit organisations (NPOs). In an age that will increasingly question the profit motive, especially when meeting the needs of the sick or vulnerable, expect NPOs to grow fast. Half their income is from government contracts and grants. NPOs will be increasingly popular, 'run by people for the benefit of people', rather than with the prime motivation of securing return on capital. NPOs will be characterised by greater rigour, professionalism, audit and evaluation. Fierce competition between agencies will equal anything in the commercial sector today. Expect huge voluntary sector growth in America, where already 60 per cent of adults give on average 200 hours of time a year.

There will also be competition between commercial and non-profit organisations, with accusations by commercial companies that NPOs are artificially undercutting prices by using unpaid voluntary labour, or by cross-subsidies using donations. The whole concept of charitable work will be called into question as many agencies find themselves more and more as simple subcontractors to government, with severe restrictions on their activity. Expect the commercial sector to win a campaign for charitable status to be abolished in the case of organisations which are purely subcontractors. As part of this shift, volunteers will increasingly question whether their act of charity is simply being used to enable government to cut costs and jobs.

Unable to deliver promises

The result of all these developments is that governments will be unable to fulfil election promises. Most traditional functions of the state have drifted out of their control. Failure to deliver has contributed to disillusionment. Add to that the smell of corruption and the loss of confidence becomes acute. Even Westminster, regarded as the mother of all parliaments, has been seen as full of sleaze with evidence of back-handers, secret favours and dishonesty. Westminster may be the least corrupt democracy in the world but the British people have lost confidence in it. An opinion poll showed the following attitudes:

Disbelieve politicians	90%
Politicians tell lies to protect themselves	90%
Ministers cannot be trusted	90%

These results are almost identical to those in Russia, where 90 per cent of the people distrust members of parliament and 88 per cent distrust the government.

Countries with a reputation for political corruption

Rightly or wrongly, many countries have governments which their own people see as corrupt. It is part of a broader picture – for example the racketeering and bribery scandal in Japan among top financial institutions, or the collapse of the Bank of Credit and Commerce International with debts of £12 billion.

Bribery is often seen as just a local tariff payable to get things done. But poor people in these countries are the losers. Corruption means good projects are squeezed out in favour of bad ones. Honest officials give up. Bribes grow bigger. It becomes all but impossible for a company or a foreign government to do business without playing the corruption game. Scruples are soon swallowed. Germany is one of several nations in the EU where until recently bribes were tax deductible, while in contrast the US has always theoretically regarded bribery as an offence. Transparency International is an anti-corruption pressure group based in Berlin. There will be more such groups, together with new 'bribe-free zones' or 'islands of integrity' with public pledges by all who operate there not to accept or pay bribes.

The slow death of democracy

What is the point of voting when you cannot believe the words of a manifesto? What is the point of listening to a television interview with the President when you can't trust what he says? Voting is going out of fashion, young people are drifting away in droves. Seven referendums were annulled in Italy in summer 1997 because too few people turned out to vote. Less than half of America bothers to vote. A recent UK 'Big Brother' TV show polled more votes than Tony Blair. Expect more countries to follow the example of Australia, where there is compulsory voting. Expect electronic voting to become a low cost and accepted method for referendums, with the same channel being used for instant opinion polls on a wide array of issues, and influencing policy.

Tribalism in political leadership

When faith in ideology and parties dies, trust in the person is all that's left. Leaders rather than policies will dominate the future and new 'tribal elders' will emerge. Emotional attachment to a person and a group will be more powerful than electoral promises.

Hence Nelson Mandela, a man held in prison as a subversive for 27 years until 1990, became a national presidential figure in South Africa. He commanded the nation to change and it did. But where do you find another Mandela for the third millennium? Trust and respect are the pillars of tomorrow's politics. People like Mandela will continue to command the world stage, and will be rewarded with international honour. They will provide the future with global leadership based not on nationality, far less on party, nor even on office, but on international recognition among billions of people that here is someone (at last) who is worth following.

But this will be an uphill battle. Fewer than 20 per cent of people in Britain think they still have leaders worthy of respect and the same trend is apparent elsewhere.

Death of political theory

The old left/right politics is dead. Old politics lumped together social justice and compassion with moral liberalism, and free markets with public morality and personal responsibility. But left-wing ideology finally died with the collapse of the Eastern bloc, while the right-wing, 'hands off' government of the 1970s and 80s fell apart when confronted by the massive social challenges created by urbanisation. Right-wing rhetoric had little to say to those experiencing a taste of living hell on high crime, inner city estates, while left-wing state control had no power to create wealth.

Throughout Europe political parties have drifted right, particularly in the south and east. Spain, Portugal, Italy and Greece are countries in which the left has moved most sharply, together with countries such as Poland, Hungary, Estonia, the Czech Republic and others. The notable exceptions are France's Socialist party and Germany's Social Democrats. And both France and Germany are in danger of being left behind by the harsh realities of globalisation.

NEW POLITICS

Party politics has been replaced by single issues

Single issues will be the most important driving force in politics for the future.

Old party politics	*New issues*
World view	Narrow agenda
Systematic	Campaign
History	New causes
Tradition	Radical
Left/Right	Sometimes irrational
Examples of single issues	
Environment	Europe/world
Animal rights	Britain
Abortion	US
Genetic revolution	Germany
Holocaust gold	US/Switzerland
Farm land rights	Brazil
Child labour	US/Europe

Single-issue politics

In a recent UK election it was highly significant that two new parties fielded between them hundreds of candidates – on single issues. One was the Referendum Party, promising a vote on Europe, and the other was the Pro-Life Party. Neither had a comprehensive manifesto, a plan for the nation as a whole. They were interested in one issue only. The trouble is that you can't run a nation for long on a cluster of single issues. If you do, you turn out to be rudderless, without any underlying purpose.

So what does 'single-issue' politics mean for voters? It means that manifestos in future are likely to be non-controversial lists of centre-of-the-road policies plus a cluster of other 'populist' single issues such as environment. Pragmatic politics means that governments are more likely to change policy or introduce one between elections, as a result of vigorous lobbying. Expect opinion polls (a poor man's referendum) to count more than parliamentary debate and more referendums on important issues.

Abortion is a big issue in the US

Abortion is an example of a growing single issue: pro-choice v pro-life. It is a quarter of a century since the famous Roe v Wade case when the Supreme Court decreed that abortion was a constitutional right. Since then there have been a number of legal restrictions. The anti-abortion movement in the US is now bigger than the civil rights movement of the 1960s. Tens of thousands have been arrested, cautioned or imprisoned, while many pro-abortionists have been threatened, assaulted or murdered. In one US state in 1996 there were two major incidents every week: a clinic bombed, a death threat, a bullet through a window. From 1977 to 1993 there were more than 1,000 reported cases of bombings, kidnaps and arson attacks.

A series of murders provoked Congress into passing the Freedom of Access to Clinic Entrances Act in 1994, with punishment of up to 10 years in prison and fines of up to $250,000. While the scale of attacks has markedly reduced, the passions remain simmering, stronger than ever. Single issues are more powerful than state law. Here was a great nation that allowed abortion but where abortion was almost outlawed in practice. In one state, at the height of the protests it was hard to find a doctor willing to conduct abortions. Then the ban came on so-called partial-birth abortion – so single-issue groups can and do alter national law. President Bush is as vulnerable as was Clinton.

Single-issue groups are growing fast

Worldwide, Greenpeace has 3 million members in 32 countries and a budget of $146 million. In Britain Friends of the Earth and Greenpeace have more members together than the Labour Party which won a landslide victory in 2001. Greenpeace is adding 10,000 UK members a year with a national budget of £4 million. It is just another single-issue group. Yet Greenpeace is fading in the US, with membership tumbling from 1.2 million in 1991 to fewer than 400,000 today. Hundreds of companies now embrace environmental concerns and there has been a reaction against strident activism. Expect more traditional groups such as the Audubon Society and the Sierra Club to continue to grow.

Environment is No 1 single issue for tomorrow

The environment will be the dominant single issue for tomorrow, feeding into post-millennial fears that the third millennium will be the last. Every now and then there is a major accident which focuses huge attention. The Union Carbide disaster killed 2,000 in India in 1984, and injured tens of thousands of others, after the release of clouds of highly toxic gas from a chemical plant.

Two years later Chernobyl spewed tons of radioactive gases and debris into the atmosphere, creating a huge radioactive no-go area in the Ukraine and dumping so much radioactivity on western Europe that in 1997 Welsh farmers in Britain were still forbidden to send lambs to market in some areas because of the risk to those eating them. Then there was the Exxon Valdez oil spill in Alaska, and others in Europe, together with global scares over BSE after cows in Britain were fed meat products infected with prions. Then there was global warning.

A single issue can wreck your business

The boards of just about every high profile company can expect to meet single-issue activists at their annual shareholders' meetings, whether Shell, Nestle, Lloyds, Costains, Esso, Texaco or Prudential. CEOs now need to know how to handle people hanging from ceilings, running naked down the aisles, shouting and asking awkward questions. Road builders, arms makers, water and oil companies are just some in the firing line, together with companies investing in operations with wasteful carbon emissions, child labour in India, animal experimentation for medical research, GM food – the list is almost endless.

Greenpeace and Shell – a sign of things to come

The row over the dumping of the Brent Spar oil rig in the North Sea was an important warning to every institution. It is a vital case history because big mistakes were made by both sides. Similar mistakes have been made by other corporations since. There was long-term damage to Shell, and great damage was done to the entire oil industry in image and increased disposal costs. There was also damage to Greenpeace, and costs to the environment and consumers. Dumping Brent Spar was always going to look bad – a large contaminated structure was to be sunk to the sea bed where it would gradually rust to pieces, releasing toxic waste. But Greenpeace got their facts wrong about the amount of oil and other contaminants on the rig. They ran a slick publicity operation and there was a vigorous consumer reaction, especially in Germany with a widespread boycott of Shell products. British Prime Minister John Major had previously said that the government would support sea dumping as safe, and a far more satisfactory means of dealing with the rig than land disposal. The government finally caved in and said they now wanted it disposed of on land. This single-issue group had become more powerful than the government.

Then Greenpeace realised their mistake. The amount of oil on board was less than feared. But the damage had been done and despite a public apology Shell was forced to continue with land disposal, even though most experts, and Greenpeace, now agreed that it was a worse solution for the environment.

It all showed the power of single-issue groups, small numbers of activists – and the key was control of the media. When it came to the raiding of Brent Spar, Greenpeace had a great advantage. TV news always gives priority to news stories with images and Greenpeace owned them all. They had cameras and satellite links on board the rig and did not allow other crews to land. They controlled the entire output.

The scenes are as spectacular as in any movie. Greenpeace chartered a helicopter to ferry supplies from a nearby vessel to the rig. Spectacular shots were taken of water jets from fire tenders actually hitting the helicopter as it hovered in mid-air hundreds of feet above. The craft shakes and whirls. Ten seconds of images like that seemed to brand Shell as the big bully and the activists as schoolboy heroes. And anyway, what did Shell have to hide? It looked dreadful and from that moment the battle was lost.

Corporate ethics is big business

Companies like Body Shop will continue to come and go, riding the wave of ethical fashions and fads with products designed to please the ethically sensitive, while other corporations will rush to get close to activist groups to gain sympathy and support.

Ethical investment funds are already 8 per cent of funds under management in the US and will continue to grow, prompting intense debates in future about definition, and regular scandals as 'ethical' companies are discovered to have broken unwritten rules. Getting listed or unlisted as an 'ethical' company will be enough to see a CEO rewarded or sacked. Of personal pension fund holders in the UK, 75 per cent now say that they would prefer their pension fund to have an ethical investment policy even if it means a lower return. In fact returns have been excellent.

All other investment groups will be forced to notice. Ethical investment is now mainstream – even when it hurts investors' pockets. And that means pressures on all large companies to prove their activities are worthy of 'ethical' investment.

A single issue can strike fast and hard, suddenly threatening to overwhelm every other board priority. Every corporation should have a team watching out. Subscribe to newsletters. Join the relevant organisations, send members of the company to annual conferences. Listen to the gossip over coffee, and to comments on debates from the floor. Early warning is absolutely vital. Every large corporation should regularly give itself a single-issue 'health check'. The key to survival will be to seize and hold the 'moral high ground' in every way, with clear corporate values and well understood systems of checks, sanctions and public punishments.

McDonalds wins libel case but loses face

Look what happened to McDonalds. The world's biggest burger chain was attacked in pamphlets that made serious accusations. The authors were an almost penniless man and woman with strong views. McDonalds decided to make an example of them – such accusations had been made in other countries by various groups. The case became the longest libel case in British history. The trouble is that libel law prevents repeat of a libel – except in a court. And what is said in court can be printed indefinitely in a report of proceedings. So the case made matters far worse. The pair were ordered to pay just £60,000 damages (which they could not afford) when the judge agreed that some accusations were true. More recently rioters attacked McDonalds in India after it was revealed that 'vegetarian' french fries were contaminated with tiny amounts of beef products.

Holocaust gold

Over fifty years after the end of the Second World War, who would have thought that Nazi gold would have hit Switzerland so hard? After major rows and accusations of a cover-up, Swiss banks relented on traditions of absolute secrecy and published a list of several thousand dormant accounts from the war years, the majority of which might well relate to those who perished in the Holocaust. Unfortunately the same mistakes were made as by Shell. Instead of a swift decision to do what they were later forced to do anyway, the banks hesitated and seemed not to care. They collectively misjudged how acutely sensitive the issue was in the US, and only began to react when their own US offices began to warn of losing business over the 'scandal'.

Trade–environment disputes will grow

The World Trade Organisation is already being asked to sort out conflicts arising from different environmental standards between nations. Examples include the US–Venezuelan row over high-sulphur oil exports. Another is the US dispute with India, Malaysia and Pakistan over the export of shrimp caught using 'environmentally harmful' fishing gear. The US has worried that Mexico's lax standards will mean companies in the US can relocate production there to cut down on environmental costs and then undercut US companies. There are already threats by some countries to slam extra import duties onto products made in countries with poor standards, to protect domestic producers.

The counter-argument is that green regulations make firms more competitive and creative, producing higher quality products for increasingly fussy markets. The theory is that these companies stay ahead, controlling pollution and degradation and winning new niche markets. In practice it is far more tempting for some industries to relocate where production is easy and cheap. Either way, environmental business is growing and is already valued at $1 trillion a year, a figure expected to rise dramatically by 2010 with carbon trading.

American mega-waste

The US Heads the World League for Waste and Consumption

- 35% of the world's cars for 4% of the world population
- 19% of the world's garbage – 3.5 lbs in weight a day per person
- Almost three quarters of a ton per person a year
- 30% of the rubbish is packaging

Every year that means:

- 15 billion disposal nappies
- 2 billion razors
- 1.7 billion pens
- 45 billion pounds of plastics
- (200 million pounds of plastics is exported to Asia and land-filled there)

In contrast Europe is set to recover 90 per cent of its packaging waste in ten years. Traditional repair shops and fix-it industries have died, but are being replaced by an increasing number of waste exchanges and re-use centres. One of the problems is that local repair labour costs are very high compared to production costs, which are low because of mechanisation and cheap labour. This is an area where every individual can take positive action, at home and at work, with purchasing, consumption and disposal decisions. It doesn't take much effort, for example, to start a paper recycling scheme at work – assuming of course that your company is still using paper.

ENVIRONMENTAL MEGA-ISSUES

Global warming

Mega-issues such as global warming and atmospheric pollution will bite hard. Global warming will certainly be a major issue for the next half century. It is beyond debate that the level of carbon dioxide is rising rapidly in the global atmosphere. World energy demands are growing at 2 per cent a year. By the year 2010 developing countries are expected to produce more greenhouse gases than developed countries. By 2020 world energy consumption is set to treble, mainly because of fast-growing Asian economies. Worldwide $3,000 billion will need to be invested to realise this. These figures assume that population increases by 50 per cent, with a doubling of both world income and energy demand. Even if these figures for energy demand are not reached by 2020, it is highly likely that they will be by 2035. All this points to one thing: unsustainability. With only 4 per cent of the world's population, America produces 20 per cent of greenhouse gases. What happens when every developing nation wants an American life-style?

The Kyoto Protocol agreed in 2001 by over 190 nations – excluding the US – to limit global carbon emissions, will create huge international tensions by 2010. Expect every new flooding disaster or unusual drought to be blamed on global warming. Expect big problems with non-compliance by governments who fail to ratify global agreements, or to abide by them. Expect righteous anger from poorer nations when they find their own industrial revolutions held back by rich nations. Expect huge global markets in carbon, with rich nations failing to cut carbon emissions and only reaching targets by persuading (and funding) poor nations to reduce theirs. Expect the US to eventually fall into line with global treaties, after threats of global isolation and violent protests. Expect big research into carbon-trapping technologies, including biotech trees which grow very fast – posing further concerns about environmental risk.

Expect a huge (reluctant) rethink about nuclear power which is the only technology with enough capacity to produce unlimited energy with zero carbon emissions. Expect suggestions by 2010 that nuclear power could be used to trap carbon, removing it from the air. But feelings about nuclear power could change in a single day following another nuclear disaster like Chernobyl.

Extremist groups will take great exception to 'jet-set' travellers using more than their fair share of the earth's resources in 'selfish' travel for business purposes, when the same tasks could be carried out using the latest communication technologies. Expect airports and passengers to be targeted by activists. Expect businesses to plant forests as part of their own 'carbon saving'.

Expect protests by industry at the green tax on jobs, with an insistence that unless environmental controls are applied consistently worldwide, those countries with greater controls will simply see jobs moved out to countries where controls are weakest. Power will always be directly proportional to a country's GNP when it comes to negotiations, so nations with big GNP will end up running the new global system. It could run very well on a few hundred global agreements.

Growing worries about sun, sand and beach holidays

The damaged ozone layer will continue to fuel tourist concerns about skin cancers such as malignant melanoma, the occurrence of which doubled in a decade. Sunlight will be blamed for an increasing number of other disorders, including cataracts and non-Hodgkin's lymphoma (a kind of cancer). Skin cosmetics will increasingly emphasise ultraviolet ray protection. Sun screens are now so strong that with their use it is almost impossible to develop a 'normal tan'.

Health warnings are severe and their implications bizarre. These 'tanning agents' contain leaflets in the packs warning that 'any tan is a sign of harmful skin damage'.

Expect brown, tanned skin to become less fashionable, with a return to paleness as a sign of sophistication as it was in the nineteenth century, when a tan was a sign that you were an outdoor labourer. Beach holidays in hot countries will become suspect in the eyes of increasing numbers of people.

Expect growth in culture holidays, exploration holidays, learning holidays, activity holidays, as more people take several short breaks a year in addition to a longer vacation. Risk, excitement and experience will open new tourist markets, including countries offering extremes of hot and cold ranging from the United Arab Emirates through to Greenland or Iceland.

The legacy of communism

Communist bloc countries became the worst polluters in the world and the cost is still rising. Azerbaijan is a small country with only 7.5 million people, yet by the beginning of 2002 over 4 million tons of extremely toxic waste had accumulated there. The Caspian Sea's unique bio-resources, including sturgeon fish, are all but destroyed – affecting four nations. During the 1980s, 15,000 tons of oil and oil products, 20,000 tons of mineral acids, 800 tons of dissolved iron and 500 tons of phenols were poured into the sea. Air pollution is terrible, along with water degradation, land erosion and salinisation.

New power structures

Expect further development of renewable power sources. In the early 2000s research on renewables is less than 30 per cent of the $5 billion spent on nuclear power research every year. Expect a huge growth in wind farming, especially in windswept nations such as Britain where up to 20 per cent of energy could be generated this way. If the same levels of subsidy for nuclear energy were applied to wind, wave and solar sources there would be a rapid acceleration in development. The US aims to have a million roofs covered by solar panels by 2010 but that will not be enough to reduce demand significantly.

Expect great debate over grand tidal schemes closing off the Wash and the Bristol Channel in the UK. The latter has the second largest tidal range in the world. Expect vast schemes to generate current in desert areas from solar panels, and new methods to carry electricity large distances with minimal power loss. Expect energy generation to happen everywhere – even on the back of lorries. Sainsbury's, the food retailer, already has an experimental refrigerated lorry whose cooling cells are powered by a roof covered with solar panels.

Expect limited experiments by 2008 with solar powered satellites (SPS), megastructures in space designed to take sunlight and convert it into a microwave beam, directed back to earth. This technology will prove expensive and environmentally worrying. Panels up to 100 square miles in size would be needed to power a major city, with fears of damage by debris from space, and dangers for all life in the path of the beam. Microwaves are in one way identical to light: just another part of the electromagnetic spectrum. The electromagnetic energy in a microwave beam will be just as intense as that in a light beam carrying the same power.

Rubbish will be burned for power when it can't be recycled. A prime target is motor tyres. The US alone throws away 250 million a year. Heating produces a thick, smelly black sludge from which other products can be made. But they can also be burned in a power station, as they are already in California.

Expect new materials to be developed that will be as revolutionary as plastics were in the mid-twentieth century. Plastics changed us all, whether nylon or polyester in clothes, plastic bags for shopping or food wrapping or the dashboards of cars. Expect remarkable new fibres for clothing and fashion industries, and ever stronger, lighter products such as carbon fibre.

Expect cars with aluminium bodies to beat rust. Expect cars with bodies of 30 to 60 per cent plastic, and other compounds, for example glass fibre. Expect car engines using ceramic technology for high efficiency, low fuel consumption and low engine wear.

Rethink on cars will accelerate

Severe restrictions on car use in major cities will become a way of life, leading to permanent controls in a bid to protect public health. Cars for short journeys will become unfashionable for a growing number. Already out-of-town shopping malls are falling from favour among town planners and governments in parts of Europe, while pedestrian precincts continue to sweep across town centres. Building more roads simply increases traffic. The nightmarish gridlock of cities such as Mexico City, Mumbai and Cairo will continue to raise questions about the future of cars.

In July 1994 people were dying in London because of smog produced mainly from exhaust fumes. That summer I flew to Norway – and could not see the ground at any point, in clear skies at 35,000 feet. People were dying in Norway too. A week later I returned and my family drove to the most remote north-west tip of the British Isles to a Scottish island, where there are prevailing westerly winds, and where the air is usually very pure after travelling 3,000 miles over the Atlantic Ocean. The air was so polluted we could hardly breathe. It was the same across most of Europe.

And what about India? The air in Mumbai and Calcutta is so foul on many days that it makes that European smog look like the freshest mountain air. Consider this: if people in China own as many cars per 100 people by 2050 as we have in the West, there will be more cars in China alone by 2050 than there were on the whole earth in 2002. If they all ran on petrol there would not be enough crude oil to keep them going, nor enough wind or fresh air to disperse the massive cloud of pollutants.

In 1993 96 per cent of Chinese car sales were to government departments. China still has only two vehicles for every 1,000 households. The total market right now is a mere 400,000 cars a year. There are 10 million in total for 1.2 billion people. That is expected to rise to 25 million by 2015 and 150 million by 2030.

On the other hand there is huge global over-capacity in car factories. If all the car firms in the world worked flat out they would produce 68 million a year (including small pick-up trucks and vans). Yet in 1997 they only made 50 million –73 per cent of capacity. In Western Europe the industry worked at 67 per cent of capacity,

Japan 50 per cent and North America 79 per cent. The situation will get worse.

The result will be a glut of very cheap cars and a big squeeze on manufacturing in America and Europe. Expect strikes, demonstrations and political action. When Renault threatened to close its Vilvorde factory in Belgium, tens of thousands marched through the streets of Brussels. We will see multiple closures and rationalisations.

In the longer term, restrictions on car usage are likely to affect shopping, rail travel and where people live. They will encourage local community life with a reversal of the trend to out-of-town superstores, and will discourage long-distance daily commuting in favour of living close to work in cities or towns, or teleworking.

Individuals can take steps to tackle their own contributions to these problems. For example my own family home is within walking distance of our children's schools, major shops and railway networks, while I telework extensively. We are aiming to become carbon-neutral, investing in new woodlands to help offset our own household emissions.

Buildings will also be targeted

Around 50 per cent of earth-warming gas emissions in big cities are from homes. Expect homes in wealthy nations to become new efficiency target areas, with ultra-efficient boilers and better insulation – by law. Only 25 per cent of emissions come from travel, the rest from industry.

Transcontinental smog

In the next two decades on current trends we will begin to see transcontinental smog: whole sections of the surface of the planet where the air on the ground is unhealthy to all and lethal to some. Smog kills by aggravating asthma, bronchitis and a host of other lung conditions and by directly increasing the risk of heart attacks through carbon monoxide exposure. In a city like Calcutta that could mean up to 25,000 extra deaths a year. Across a whole region or continent the results are unimaginable.

A recent hint of what is to come occurred in September 1997 when dense smog developed over a million square miles of South East Asia. In Singapore the air quality fell to the point where breathing became the equivalent of smoking 20 cigarettes a day. President Suharto of Indonesia was forced to apologise for the forest burning, clearing land to grow food, which had been a large contributor, while Malaysia declared the severe haze a national disaster and schools were shut throughout Sarawak. The clearance itself was a product of land shortages and increasing population. Some estimates suggested that the fires released as much carbon dioxide as Europe produces in a whole year. Car exhausts were also a contributory factor.

The total population affected in some way was at least 300 million. If 50 million were severely affected, the increased deaths could have been higher than 1,000 every week, or at least 10,000 people. To this must be added the side who survived. And that is just the beginning.

But what about the impact of a larger, longer smog, continuously circulating over the whole of China, Cambodia, India, the Philippines, Australia, New Zealand, the west coast of the US, Mexico and parts of Europe? Tens of thousands of deaths, and projections that within a further 15 years such smogs will have quadrupled in frequency.

We are gassing ourselves: human beings are profoundly altering the balance of gases in the atmosphere. At first, developing nations will continue to sacrifice air quality for economic growth. But within this generation pollution could become economically damaging, not least because of a loss of confidence among institutional investors from other nations. The pressure will be to clean up, to restore the image of 'civilised' urban life.

Carbon taxes will cause international tension and conflict, even wars, unless applied fairly – but that will be almost impossible. In order to be fair, carbon rations would have to be a fixed amount per person per year, perhaps sold on the open market as a tax on all carbon consumption, with subsidies for the poorest and most vulnerable. But that means villagers can no longer cut their own trees for fuel, and will condemn the poorest nations to a relatively carbon-free existence forever. Meanwhile the wealthiest nations will continue to ransack the earth's limited carbon supplies, and grab the largest slice of all the permits to pollute the atmosphere that every other nation also has to breathe. So while America burns fuel in its cars and air conditioning, Bangladesh will drown in ever-worsening floods. Wealthy nations will boast that they are cutting the growth in total consumption, but will remain the greatest users of limited environmental resources. The loss of rain forests will add to the long-term effect, making future environmental corrections even more difficult.

Weather predictions will be more accurate

If we can't control the atmosphere easily then at least we can try to predict its behaviour more effectively. Expect medium- and long-range forecasts to increase steadily in accuracy. The limiting factor is not processing power or speed, but the lack of data sets for comparison. Accurate records of such things as sea temperature, air pressure and temperature have only been kept for a few decades, a mere blip in earth time. It will take another decade of earth-watching before we see major advances in forecasting. Nevertheless, global warming should lead us to expect greater instability in weather patterns.

Discovery of low cost clean fuels

The only clean fuel we know is hydrogen, because the only exhaust fume it produces is steam. Hydrogen buses are here already in Chicago and Vancouver, while Daimler-Benz and Ballard Power Systems are investing $300 million in fuel cell technology. But hydrogen is a low energy gas and has to be made by splitting water. The burning of fuels for anything other than heating is in any case extremely wasteful because so much is lost in heat energy. The most efficient power source for cars would be the use of hydrogen to drive electric batteries. That sounds fine for the atmosphere, but it is not. After all, where does the hydrogen come from?

The answer is that hydrogen is made in power stations by burning coal, gas or oil and generating electricity, used to convert water into hydrogen and oxygen. Sunlight can be used to generate electricity to make hydrogen, but there's not enough of it unless we blanket every desert with solar cells – which would itself alter the climate dramatically. Hydroelectric power is limited and wind turbines will never satisfy tomorrow's world.

Nuclear power is a dead industry that will revive

A graph of new nuclear power station starts show that after peaking in the late 1980s it then dropped to almost zero. What that means is that by 2030, on current trends there will hardly be an active nuclear reactor left for industrial power generation. Expect to see another Chernobyl-type nuclear disaster sometime over the first three decades of the twenty-first century as all these stations age and become less reliable. Also expect to see growing concern about what to do with all the dangerous waste. The biggest nuclear-related industry in the decades to come will be disposal and dumping.

Nuclear hazards will worry third millennialists, who will be often reminded that plutonium is 30,000 times as dangerous as cyanide, with just one particle capable of causing cancer if it enters the lungs. Despite this, pressures to reduce carbon use will make nuclear energy very attractive to some.

The Nuclear Timeline

- 1932 Atom split in London
- 1947 Britain builds first nuclear reactor
- 1958 Fusion achieved
- 1958 Secret Soviet blast kills hundreds
- 1977 US halts breeder reactors
- 1979 Three Mile Island atomic leak – risk of core meltdown
- 1979 65,000 protest in Washington
- 1981 Indian Point (USA) shut down after leak
- 1986 Chernobyl accident releases radiation across Europe

Just forty years from the beginning to the end. The love affair with nuclear power started in 1947 and ended with Chernobyl.

Expect progress in nuclear 'cold fusion' experiments and other new discoveries. These will renew the nuclear controversy in an increasingly energy-hungry world.

Co-generation of electricity

A huge, wasted by-product of electricity generation is heat, which is particularly wasteful when you consider that electricity is often used to generate heat. Expect to see many more small-scale generators, supplying heat to large buildings or housing estates as well as electricity, some of which will be surplus to need and sold to the national grid at prices which undercut big power stations.

Growing worries about electromagnetic fields and noise

Expect growing concerns about the lifetime effect of low dose exposure to electromagnetic radiation from overhead power lines, mobile phones and other devices. There have already been suggestions that there may be a link with cancer. Expect further evidence that mobile phone radiation affects brain function, as well as affecting the function of other cells – and legal action – even though risks to individuals from normal use will remain very low.

Noise is a pollutant and anti-noise devices will be developed further. Anti-noise reproduces the exact opposite of every sound wave to wipe out sound. Expect anti-noise technology as standard inside cars and aircraft, and in homes near noisy sites.

Recycling a way of life

Every year more treaties are signed on the environment – there are over 100 to date, and they are generally effective. For example, CFC production has all but ceased following global agreement to protect the ozone layer. And recycling has become a way of life in many countries, so that many newspapers now boast that over 40 per cent of the paper they use for printing is recycled.

Expect recycling to become a specialised multi-billion dollar industry assisted by subsidies and public goodwill. There will be a proliferation of waste sorting in many countries along with glass, plastic, cardboard, steel, aluminium and newsprint containers in car parks and shopping areas. Domestic refuse collection will become a new technology area with the packing and sorting of different waste types. Expect car breaker's yards to become total recovery areas, with regulations demanding zero waste.

Waste disposal companies will be popular with 'ethical' investment funds and will be profitable, mainly because of millions of free hours of labour provided by a willing population, who will clean, sort and carry billions of items every year, 'doing their bit' for the environment.

War and the environment

The trouble is that conflicts drive massive holes in all environmental programmes, as seen in the Iraqi invasion of Kuwait. Not only was there catastrophic environmental pollution through the bombing and burning of oil installations, but also continued environmental damage after the war ended. US bombing targeted desalination plants, sewerage, heavy treatment plants, irrigation and drainage systems as well as more obvious military goals. The destruction of power stations and factories led to more burning of wood as fuel.

THE GENETIC REVOLUTION

Another single but far-reaching issue is biotechnology. We are going to hear a lot more about genes in the future. We think of the computer revolution as highly significant but biotech is in many ways far more important. Computers may alter the way we live but biotech has the power to alter the very basis of life on earth. It is also a very efficient method of producing highly complex substances, whether biological compounds, medication or chemicals for manufacturing. As we have seen, health is one of the biggest areas of spending in developed countries, accounting for up to 18 per cent of GDP, and genetics will dominate health. In comparison nanotechnology will be very disappointing, except in computing.

The genetic revolution could produce a rapid increase in life expectancy. Many leading biotech researchers believe we will be able to slow the biological clock. This has already been seen in nematode worms who live twice as long when the DAF 2 gene is turned off.

Cut and splice

We now have the technology to take genes from any organism and put them into another – just to see what happens. Human genes have been added to mice, cows, sheep, rabbits, rats and fish, to name just a few. The Human Genome Project has already mapped in detail the entire human genetic code. The result will be new tests for many of the 4,000 genetic diseases and thousands of others where genes are an important co-factor.

Individuals will find gene screening becomes a routine part of life insurance and pension assessment, while some companies will try to screen the genes of certain job applicants. For example, there is a gene that makes it more likely that workers will develop lung disease if exposed to high levels of dust in the air. Far cheaper to test and eliminate carriers than to clean up the factory.

A mouse has been born with a hundred human genes (around 0.1 per cent of the human genome) while herds of humanised pigs have been created to provide humans with new hearts or other organs – probably unsafe due to the risks of transferring pig viruses to humans. Adult stem cells from bone marrow have already been used to create liver, kidney, brain and other cells. Similar work is being done using embryo stem cells – far more controversial. Expect adult stem cell technology to provide a miracle for those with diabetes. Replacement cartilage will produce perfect repairs in arthritis sufferers: scientists have already made perfect tiny human kidneys, by implanting primitive cells from 6–8 week old foetuses into mice. If they had used pigs, the result could have been a transplantable organ, but raising many ethical questions.

Scorpion poison genes have been added to cabbages to kill caterpillars – but what might they do to people? Potatoes have been made with an insecticide gene from a soil microbe – immune to Colorado beetle. New kinds of maize and other crops are on sale, some producing fungicide or insecticide in their own sap. These 'green' crops need no sprays, but are they green inside? Then there are non-bruising tomatoes and bananas which will soon contain vaccines or other medical products.

Green products growing fast

In 1996 US farmers planted one million acres with genetically modified soybeans, and 10 million acres in 1997. These soybeans were deliberately mixed with 'natural' soybeans and have been used globally in food without proper labelling. It is impossible for consumers in the US and much of Europe to choose not to eat genetically modified soya, unless they boycott all soya products.

Expect gene research to produce ultra-fast growing trees that cannot reproduce except by cuttings taken in a nursery. They will be the answer, in part, to rising carbon dioxide. The ideal tree for growth and utility will have dark green, dense foliage on very short thin branches (less waste), with tall straight trunks. Such forests will be very similar to fir forests today. No chance of any vegetation beneath. Nor wildlife.

Expect ultra-slow growing hedges and other plants, that can be triggered into normal growth by the addition of special fertilisers when required, saving gardeners the hassle of trimming or pruning. Expect species recovery programmes in zoos for extinct or almost extinct animals, or for rare animals in captivity who will not breed. Original genes from all kinds of sources will be mixed where necessary with those from similar species to recover as much as possible of the original. Sources will include frozen tissue samples and cell cultures.

Also expect to see new smart drugs, some produced by genetic engineering, that hold the promise of enhancing brain function, memory and processing speed. Such drugs are already in human trials and some studies show extraordinary improvements. Deprenyl is an example, not only possibly helping those with Alzheimer's but also extending the lifespan of rats by up to 40 per cent. Expect to see abuse of such drugs by students working for exams and others under creative pressure, in much the same way as anabolic steroids are abused by athletes.

Expect the rapid adaptation of many kinds of wildlife to the megacity, including the emergence of new super-rats as large as small cats. However these will be part of a process of natural selection, not the result of laboratory experiments. Designer animals are already here. Over a million mutants were made in UK laboratories alone during 2000–2002, each of which was a unique mix of two, three or more different species, for example, transgenic sheep pro-grammed to produce human substances in their milk. Next will be attempts to humanise cows to produce low fat milk. The ultimate goal will be cows that produce human breast milk.

It all raises huge questions of safety and ethics – including issues of animal welfare. For example one set of humanised pigs grew fast, with low fat meat, but were blind, impotent and suffered from severe arthritis, so that they could hardly stand. Some people will refuse to eat humanised animals, others will be unhappy about culling them for organs. Genetic engineering poses profoundly challenging, unanswered questions.

Biotech industry blows up and settles down

Most smaller biotech companies will struggle to justify the huge investments made as the century turns, with false hopes raised by sensational headlines. Many will fold, merge or be taken over by traditional drug companies anxious to 'buy in' expertise, while others will survive only as subcontractors. Nevertheless, huge for-tunes will reward companies that establish unique biotech products with a clear application to health, industry or food production. Expect spectacular successes like Amgen, which grew from nothing to a $17 billion valuation in a decade. Any optimism must be tem-pered by the fact that only 20 per cent of all drug trials result in marketable products, and that gene therapy can have serious side effects resulting in lawsuits.

Human cloning

Then along came cloning. Animal cloning has been possible for a long time, having first been conducted in frogs in the 1950s. Mammal cloning is relatively recent. The first method is artificial twinning: up to 128 identikit rabbits can be produced in one go this way, but not without risks. When the technique was tried on cows the calves had to be delivered by caesarean section because they grew to twice their normal size before birth.

Twinning is easy: just get a ball of cells shortly after an egg is fertilised and use a probe to separate them. If you do this early enough you find that each separated cell goes on to produce a complete new animal. Most mammals have been cloned this way. Scientists claimed (as they always do) that this would never work in humans. What nonsense: with 2,000 pairs of identical twins born every day worldwide it was clear that such cloning would work extremely well in humans, with a good safety record for producing healthy babies.

In the 1980s I met a leading British scientist who claimed he had already succeeded in making the first human clones, although they did not survive long. He was hoping to use them to make spare parts for tomorrow's people. He hoped that by implanting one and freezing the other, the older twin could use the frozen brother or sister as a factory. If the older twin needed, for example, bone marrow, the younger twin would be defrosted, implanted and culled at, say, 24 weeks. Alternatively the twin could be born and be used as a donor then, without risk to life.

Such ideas may seem bizarre but Dr Jerry Hall in Washington announced in 1993 that he too had succeeded in cloning human embryos by twinning. Then one leading expert on ethics in Britain said that cloning for spares could be quite a good idea, so long as the foetus was culled at an early enough stage. This amoral stance was just an example of what was to come.

More human clones

Then came Dolly: a cloned sheep made using a very different technique. Genes from an adult sheep cell were combined with an unfertilised sheep's egg. The result was the birth of a lamb that was an identical twin of the adult. Within 48 hours President Clinton announced an emergency enquiry, worried that the methods would be used on humans, and other governments also reacted with alarm. Clinton proposed a ban on clones being born but declared that nuclear transfer experiments could continue. In other words clones could be made but must then be destroyed. I predict therefore that unless American laws are tightened, human clones will be made in the US, or using techniques refined in the US, but will be born elsewhere. Some scientists are saying they have already created viable cloned embryos. Births are only a matter of time, and may already have taken place, if current claims are to be believed.

Market for clones

There is certainly a big market out there. I have a constant stream of enquiries to my website from people asking me if they can be cloned even though I am strongly against it. 'Diane' told me that she wanted to clone her dad who was dying, offering her own womb as a surrogate: 'I intend to see that he goes on in this world.' She wanted to give birth to her own father.

Another woman suffering from infertility wanted to use her own cells to make a baby rather than use donated sperm and eggs. A student wrote that 'it would be so neat' to be cloned. Six per cent of the US population thinks that human cloning is acceptable. However birth of the first clones will cause a backlash in growing numbers who already feel science is drifting out of control.

Cloning will be very popular among some groups with wealth: the ultimate in pedigree children. Supermodels could make a lot of money selling cells from their bodies to cloning merchants, who would offer childless couples the child of their dreams by creating a clone and then implanting it in a surrogate. In future humanised apes could be used as surrogates. Scientists are already able to sustain a foetus in late development inside a completely artificial womb.

At a day-to-day level, cloning technology will mean that infertile couples can have a twin of the father or mother as their newborn baby, or that parents can 'recreate' a dead child. These will be the justifications used for pursuing the technology. However there are enormous safety and psychological risks for the child. Even if the child is healthy, what will be the emotional impact of growing up knowing that you are your mother's or father's twin? What about the pressures from a parent to 'relive' their own genetic potential – for example to see how musical he or she might have been if given music lessons?

Designer babies

We already have the technology to make children to order, using the same technology as tried in animals. Physical and mental perfection is a dream for many. For those whose genes are already fixed, plastic surgery will continue to offer remarkable remoulding of faces, ears, necks, breasts, buttocks and thighs and will become increasingly common as a death-defying generation attempts to stop the ageing process.

Cloning raises interesting possibilities: women no longer need men and can produce an entirely female society. So for example, a single woman, using her own egg and a skin cell, could give birth to her own twin, and in due course that twin could clone herself, with the process repeated through many totally female generations. Or we could create an entirely male society, once animals have been humanised sufficiently to carry humans in their wombs.

Cloning the dead will also be possible. Dolly was cloned using frozen cells, so any human could in theory be recovered from the grave as a baby so long as cells had been suitably frozen before death or shortly after. Living cells can be found in the human body undamaged for up to a week, so cell removal could be delayed for some time after death. Another way to clone the dead is from cells grown in culture rather than frozen. Such cultures can be maintained indefinitely. This means that a child dying of cancer could be 'recovered', allowing parents to give birth to an identical twin.

Of course the easiest way to alter the human race is the oldest method of all: mass sterilisation or genocide of the undesirable. Some 60,000 forced sterilisations of women with 'unwanted mental and physical characteristics' were carried out in Sweden from 1935 to 1976, with similar practices on a smaller scale in Denmark, Norway, Finland and Switzerland. Meanwhile 20 per cent of the entire world's population is already banned by law from having children if the state decides that their genes are not worth reproducing – in China. Far less draconian but just as significant will be the widespread testing and destruction of embryos or foetuses because their parents decide that their genes are 'sub-optimal'. Such testing is already routine in IVF clinics.

So what does all this mean for the longer-term future? The third millennium will see human beings begin to take over life itself, redesigning plants, trees, vegetables, animals and even themselves. Every day new genes are discovered and more is understood about what each does. By 2010 we will have a very good idea about what will happen if particular genes are switched off and on. The twenty-first century will be known as the age of the gene.

Genes are the ultimate in miniaturisation. A conventional laboratory to make insulin would occupy a vast area and cost several billion to construct, needing huge numbers of staff. Yet that entire production unit can be compressed into not just the size of a house, not just one room, not merely a single flask or test-tube, but into the cytoplasm of a single living cell.

Once a single bacterium receives the human gene for insulin, it carries on dividing and growing forever, eating food and making insulin. Insulin production becomes as simple as brewing beer, except that we are using bacteria instead of yeast and the product is insulin instead of alcohol. Every complex chemical product you can think of will be made by gene technology in brewing vats. Medicines, vaccines, precursors of new plastics, new fuels – whatever. More complex substances can be made in genetically engineered insects or in the milk of mammals such as cows and sheep.

And, of course, whole new organs too, although these will need to be made inside humanoid bodies with heart, lungs, liver and kidneys. In future we will see human beings created somewhere in the world which are growing, but are technically dead because they have no brain, possibly no arms or legs, created solely as organ factories. Expect many ethical questions about gene technology to be dominated by the 'yuk' factor, which will determine not whether something is right (which is too confusing to think about) but merely whether it is acceptable to the majority of people.

Bio-computers

Biotech will be used in new generations of intelligent machines and computer chips will be connected to living tissue. People with nerve damage will have computer-enhanced muscle control, or improved sight or hearing. In the future we will also be able to connect chips direct to the human brain, as has been done in mice, perhaps enhancing processing power and memory as well as communication

– allowing, for example, the direct reception of data transmitted through radio waves.

Humonkeys

It will not be long before humonkeys have been made. Perhaps such embryos already exist. The technology is proven. Hybrids are easy to create. Take geep for example, a combination of sheep and goat, made by rolling together two balls of cells from two different embryos shortly after fertilisation.

But how many human genes does an animal have to have to gain human rights? A lawyer friend of mine says that the critical factor would be a creature with more than 50 per cent human genes, but that is incorrect. We only differ genetically from monkeys by 3 per cent and from amoeba by around 14 per cent. So if you are adding 1.5 per cent of human genes to a monkey cell from which a clone will be made you had better watch out. A mere 0.3 per cent of human genes put into a monkey could be more than enough to give the monkey speech.

Can monkeys go to heaven?

Theologians, philosophers and lawyers need to think now what their reaction will be when such a hybrid is displayed to the world, as it most surely will be. Is it a monster to be destroyed? Does it have a human right to life? Can it be eaten? Is it morally responsible before a court of law? Can it be tried for murder? Is it allowed to marry and procreate with 'normal humans' or to mate with other animals? Is it in need of salvation? Does it have a soul?

And long before the headline of its existence reaches us you will find religious leaders are being confronted by new forces, which will be added to by the animal rights movement as it seeks to blur the ethical distinction between animals and humans. It could all produce a crisis of faith for many, brought up on the traditional teaching that humans have been created 'in God's image'. So what is that image? Are monkeys 97 per cent of the image of God? Is all life a manifestation of the image of God to some degree or another?

These questions may hit us far sooner than we think. Historically a key trend can be seen when it comes to biotech: unlike computers, where industry experts hype the next steps and where most people have some understanding of the speed of progress, biotech is deliberately downplayed by those who know most. The atmosphere is one of secrecy and flat denial, for one reason: fear. Computers raise no great moral challenge and do not threaten the safety of life on earth. Biotech does. Few health and safety issues are raised by computer production and no ethical issues, yet such things happen every day in biotech.

Computer developers are unhindered in their work. They can enjoy the intellectual challenge of pushing technology to the limits without worrying about prosecution or being slammed in the media. Biotech specialists are quite different. They are sensitive to the delicate nature of their work. They fear uninformed public reaction and shun the limelight. They keep quiet about experiments and sometimes (for example the British cloning scientist in the 1980s) do not even publish at all. Ethical committees tend to be dominated by those in the industry who are biased towards too little regulation rather than too much.

Germ warfare

We now have the technology to create highly dangerous human viruses for research into disease or for use in war. By 2010 we could have the capability to create viruses which selectively target specific organs or racial groups. This high-tech ethnic cleansing will cause international outrage but will be difficult to stop. The poor man's ethnic cleansing machine could be a child vaccination programme, which amongst other things contains a virus with an outside coating similar to that of human sperm. The result would be antibody formation against sperm, and infertility in a future generation of men. Expect intense efforts to reduce or stop global biotech warfare research.

Islam takes a stand

Four hundred Muslim intellectuals gathered recently at Jakarta. In a joint statement they declared that revolutionary changes in science and technology had 'reduced man to a material being that is spiritually bankrupt, morally unbound'.

This anti-science feeling is general and widespread. Of British adults, 83 per cent say that modern science creates as many problems as it solves. Yet 81 per cent also think we are fortunate to live in an age when scientific development is proceeding at such a pace. The logical consequence of the first finding is that spending on science does not improve quality of life any more than it damages it. What, then, is the point of previously much-worshipped scientific progress? This is a fundamental shift from mid to late twentieth-century optimism, to the first signs of a third millennial rejection of the logical and rational.

Thus scientists are increasingly a race apart, drawn into their own world by the eccentric belief that science means benefit to people. Many scientists are baffled and perplexed by what they see as the negative irrationality of so many ignorant and prejudiced people who would want to wreck their work, given half the chance.

Scientists already find themselves increasingly on the defensive, retreating into the comfort of professional forums and conferences, bastions of intellectual openness and inquiry. This is why so few research scientists today are comfortable with ethical committees controlled by lay people who 'just don't understand'.

Expect technology and foreign business to be blamed

Expect many people in emerging nations to reach saturation point in their willingness to accept new technologies and foreign intervention. With increasing social dislocation and unemployment, many will question whether more technology and more foreign investment is healthy for a nation.

Who owns a species?

The world will soon have to face more big questions arising from biotech. Is it right to allow a company to own an entire species? Is it right to create a species which by its genes is guaranteed to suffer? Both questions have been raised by the creation of the oncomouse, designed to develop fatal cancer 90 days after birth. The oncomouse was created in America for the testing of cancer treatments and is commercially owned, protected by patent.

Patents on human genes

Is it right for companies to own human genes? A man in the US developed cancer and gave cells for research. The genes were used to develop a diagnostic test and the process was patented. He was furious. 'I own my own genes,' he said. He challenged the company and fought them all the way to the Supreme Court. He lost his case. Humans no longer have the right to own their own genes in the US.

We urgently need gene technology to feed the world and prevent disease – but we do need to ask what kind of world we are creating, now we have the ability to alter the very basis of life itself.

ANOTHER 'BIG IDEA'

Clusters of single issues do not make a political creed. The vacuum in politics will remain. Ever since communism collapsed there has been a void. Communism defined everything. So long as the Eastern bloc remained, the rest of the political framework made sense. But where are idealism and energy in political life today? Nowhere. And in the depths of this empty chasm we will see new things emerge.

Lesson from Communism

Karl Marx was born in Prussia in 1818 and died in London in 1883. He wrote *The Communist Manifesto* in 1847, and it was published in 1848. Marx took ideas and applied them in a programme for the whole of humankind. He was a product of his time: a protester against the Industrial Revolution, which placed so much wealth and power in the hands of so few, enslaving millions in primitive working conditions.

Communism was driven by a cluster of single issues, such as

over-industrialisation, worker control, equality of wealth. Yet the communist revolution only began more than 50 years later. Lenin was born in 1870 and formulated his own version of Marxist thinking in the early years of the twentieth century, publishing *What is to be Done?* in 1902. The 'Big Idea' came after the death of the original thinker.

Tomorrow's 'Big Idea'

Expect to see another 'Big Idea' emerge, radically different from any large-scale political system seen in the twentieth century. This new 'ism' will feed on the stored-up energy and desire for change created by four twentieth-century revolutions: Information, Communication, Automation and Globalisation. The longer the delay in its coming, the longer and deeper the vacuum will have become and the greater the speed with which it is likely to grip the earth. The four twentieth-century revolutions themselves will guarantee that when the new 'Big Idea' arrives, it will impact on politics and government action at tremendous speed.

What Will Be This New 'Big Idea'?

Listen to the voices of single-issue activists today and you begin to get an idea of some of the elements that will be swept up in the 'Big Idea'. It is likely to be:

- driven out of a set of writings
- backed by a charismatic personality or personalities
- able to catch the popular mood
- deeply satisfying to millions who have felt lacking in direction
- capable of mobilising nations and armies
- the spirit of the new age in the third millennium
- radically different from the old left/right
- a mass movement hard to analyse or describe
- constantly adapting and reinterpreting
- rapidly changing
- long lasting in its effects
- highly confusing to old 'logic' politicians.

What happens if you roll post-modernism, new age and organised religion together – all of which are growing globally? You get the beginnings of a new world order. Global government for a global village. In this ever-shrinking global village where sovereign states are weakened by global forces, it is unthinkable that traditional governments will survive.

CHALLENGES TO MANAGEMENT

Ready for reaction against twentieth-century values
◆ How ready are you for a significant third millennial shift in market 'culture', in the way people think and behave – for example the possibility that sun-soaked holidays may go out of fashion?

Loss of government power
◆ Are you as well 'in' with the EU in Brussels as you have always been with New York, Washington, Bonn, London or Paris?
◆ Does your company influence EU policy formation or similar regional trading blocs elsewhere?
◆ Are you integrated into other networks such as the World Trade Organisation or United Nations?

Influence of non-profit organisations
◆ Are you ready for non-profit organisations to sweep beneath you with lower prices, no shareholder dividends to find and with voluntary sector subsidies, including tax advantages?
◆ Are you prepared for the whole concept of 'reasonable profit' to become high profile, with large profits increasingly questioned as anti-social?
◆ Does your company have an adequate community action pro-gramme to help offset some of these pressures?

Bribery and corruption
◆ What is the official and unofficial line regarding the paying of bribes or backhanders to get things moving in countries where such practices are considered normal?

- ◆ Do all staff adhere to your policy?
- ◆ Does the policy or practice need review in the light of the rapidly changing climate of public opinion in wealthy nations?
- ◆ Are you prepared for current practice to be mercilessly exposed by competitors or others in the global media?

Single-issue activism
- ◆ Do you have an effective way of monitoring single issues relating to your work?
- ◆ What early warning system do you have in place?
- ◆ Do you have positional statements worked out which can be released at very short notice on a wide range of single-issue time-bombs, any one of which could attract overwhelming negative public attention at very short notice?
- ◆ Do you have a think-tank which applies up-and-coming activist issues to current corporate activity?
- ◆ Do you have a rapid-response media unit, able to make an instant, confident, well-considered response to a breaking news story, within ten to twenty minutes, 24 hours a day, representing the whole company in an authoritative way?
- ◆ Have you ever carried out a 'practice run' responding to a major story?
- ◆ When did you last do so?
- ◆ Do you have an ISDN radio studio in-house?
- ◆ Is it where it needs to be?
- ◆ Do you need one in each divisional/regional HQ?

Single issues into corporate policies
- ◆ Have single issues been adequately reflected in corporate policy, e.g. energy use, waste recycling, environmental degradation, smoking, ageism, racism, sexual harassment, ethical investment, genetic screening, animal rights?

PERSONAL CHALLENGES

Single issues
◆ What single issues are most important to you?
◆ What steps can you take as an individual to pursue that agenda at home or work, e.g. care for the environment?
◆ What steps can you take to recycle waste?
◆ What steps can you take to make your lifestyle less car-dependent?

Genetic revolution
◆ If your insurance company insists on gene screening, are you sure that you want the information for yourself?
◆ How will the knowledge affect your attitude to life?
◆ Do you want to eat genetically modified food?
◆ If not, are you looking out for labels – where they exist?

CHAPTER 6

Ethical

A new morality

We have seen a world that is increasingly Fast, Urban, Tribal, Universal and Radical – but what does it all do to people? Is this really the kind of world we want to live in? And where do we decide what is right and wrong? As we have seen, whenever we think about future trends we are forced to consider these 'softer' issues again in the light of them.

THE FINAL FACE OF THE FUTURE IS ETHICAL

The final face of the future therefore is Ethical: to do with who we are and what we want to be, how we should behave, our values and beliefs. In some ways this face is the most important. It is central to our being. It is the answer to many concerns about the future raised by earlier chapters. It is the balance to all that has gone before. Our values carry us through times of tremendous change when the whole world appears to be endlessly spinning. They provide context and meaning.

Our values are the bedrock of an urbanised society, providing answers to the problems of social decay, relationship breakdown and addiction. These same values can turn the vicious tribalism of civil war and other conflicts into forces for good, for stability, for belonging. Values provide the framework for constructive globalisation, for new political thinking and for new scientific horizons. Values are the basis for all individual life and for all community. Without personal values we become robotic, instinctive creatures with no sense of meaning, purpose, direction or morality. Without common values social interaction, community life, communication and commercial activity become all but impossible. Values define us, they provide the framework by which society operates. Personal and community values often differ from corporate or globalised values, driven often by a far narrower agenda such as return on capital or corporate survival. And values are often forged through defining moments.

Defining moments

Defining moments are points of no return in history: 'Never again'. Defining moments happen when nations react to a trauma. Their effect lasts a generation. They define the ethical values of that generation. With the increasing sophistication of communications and greater access to traumatic news events, there will be more and more defining moments beyond the millennium. Our new ethical code will be based on lessons from these defining moments.

The Second World War was a defining moment. Never again must there be a world war. Never again must we see a nuclear weapon exploded in anger. Never again must we see millions of people herded into death camps.

The Vietnam War was also a defining moment. Around 50,000 US citizens were killed. A generation later the shadow of Vietnam hangs over every American foreign policy decision. It dominated the distancing of the US from involvement in Bosnia and Kosovo.

The terrorist attacks on the Pentagon and World Trade Center twin towers on 11 September 2001 was another defining moment, not just for America, but for many other nations. It was a wake up call. A defining moment was the global unity in the UN about the desirability of disarming Sadam's Iraq, and the profound disagreement about how it should be done. Yet another was the collapse of Enron and the epidemic of corporate scandals, leading to a complete rethink about corporate governance, corporate social responsibility, share options, excessive remuneration packages, conflicts of interest, transparency, trust and so on.

Laws define ethics

The passing of laws can also become 'defining moments' and define ethics, as well as being an expression of them. For a world which has largely rejected absolutes, laws are the way we make sense of the grey area in between. Yet the legal system itself, in some nations, is a mess – particularly in the US.

US legal suits gone mad

Expect a crisis in the US legal system. Comprehensive reforms will be under way by 2010. Every aspect of US life involves lawyers in a way which amuses and shocks those in other nations. Lawsuits have tripled in 30 years. Laws cost money, as do civil actions. The US will price itself out of the world market in key industries without radical changes.

Take employment law. Why bother to base a business in the US if it means risking expensive litigation? Already five out of six corporate executives say that fear of lawsuits increasingly affects their decisions. The number of discrimination cases at federal courts doubled between 1993 and 1996 to 23,000. The number of attorneys in the field tripled. Group cases bring big rewards for law firms. When Publix, the employee-owned supermarket chain, settled a sex-discrimination class action on behalf of 140,000 employees for $81.5 million, the legal firm won $18 million.

Expect more mediation and arbitration, price wars, competition from non-lawyers for some services, judges rather than juries setting damages and new limits on civil actions. Consumer activism will make it likely that other nations will move closer to the US system while the US carves out reforms.

Single issues and ethics go hand in hand

Single issues define the problem, but ethics tell you what position to take. Expect more fierce debates and soul-searching over such issues as arms sales, as attempts are made to define exactly what arms are. Do you include machine tools used to make arms, for example? A globalised company can find itself with several conflicting positions on such an issue: shareholder values, public perceptions in manufacturing and buying nations and the opinions of company employees. There may also be a variety of government attitudes, even within the same generation – ranging from approval, to turning a blind eye to outright opposition and public prosecution.

Political correctness and thought control

Political correctness will grow in power in the next millennium as single-issue groups attempt to control the words we use. It is hard to express certain ideas if many words are banned. 'Mentally challenged' instead of 'mentally handicapped'. 'Senior citizens' instead of the elderly or retired. 'Visually impaired' instead of blind. There will be increasing conflicts between those campaigning against discrimination, who want everyone to be seen as the same, and other activists who want to draw maximum attention and sympathy to particular groups for fundraising purposes.

Countless charities today are faced with a stark choice: be politically correct and broke or sensationally incorrect and full of funds. So a cancer charity, to be politically correct, should refer to 'people with cancer', rather than to 'cancer sufferers', because the latter implies that all people with cancer suffer. Yet suffering is what triggers sympathy.

Defending civil liberties

People today can have their homes broken into by the police or secret service, their telephone tapped and bedroom or office bugged, simply because a politician or senior policeman says it should be done. No warrant is needed in many countries: the police can hold you for seven days without warrant or explanation. You can be arrested for joining a peaceful demonstration. If you are silent on arrest a jury may be invited to conclude that you had something to hide. You have no right to information the government holds about you. Britain and the US have a benign democracy, but in a country with a malignant dictator laws like these would be thoroughly oppressive.

Expect to see civil liberties on the agenda of most nations of the world, especially as electronic networks become more powerful ways of tracking people. Human rights will continue to be a major issue, especially in the negotiation of trade agreements with developing countries. There will be many agonies of conscience over whether a government should buy or sell in major contracts with 'odious' regimes – and debates on how such things should be determined.

Expect further human rights challenges by the US of countries like China to continue to fall on deaf ears until the early years of the third millennium. China believes the greatest human right is to be able to eat food rather than starve, and give citizens a reasonable standard of living, and believes it is being successful, while being more open than for decades as an evolving society.

Human rights and human responsibilities

Expect to hear much more about human responsibilities, with responsibilities and rights becoming equal pillars of global codes of ethics. The Universal Declaration of Human Rights was a product of the Second World War, and was set out in 1948. Expect a similar Declaration of Human Responsibilities. The InterAction Council, a group of international statesmen committed to global responsibility, said recently: 'In a world transformed by globalisation, common ethical standards for living together have become an imperative, not only for individual behaviour but also for corporations, political authorities and nations.'

A new motivation

While many struggle to survive, with both parents out at work, and adults labouring almost until the day they die, growing numbers in many other countries have become so well off that they no longer need to work to eat. In Britain 55 per cent of 40 year olds want to work fewer hours for less money. Most of them rate family time higher than income. This is a huge shift in attitudes from the Thatcherite 1980s, and similar trends are growing in the US. An example was the decision by one of America's highest-flying women executives to quit her rumoured $2 million a year job as president and CEO of Pepsi to concentrate on being a mother – at the age of 43. Three years earlier the UK head of Coca-Cola had done the same. Fifty year olds are taking generous early retirement packages, and are taking on modestly rewarded charitable work. Many 40-year-old high fliers have made enough to stop working, if they slim down consumption and live in a smaller house. Downwards mobility is increasingly common and will become more so. Fewer hours, less ambition, more personal rewards and new priorities, as seen in the boom of community volunteering.

Reaction against speed and constant change

Speed and constant change will give added value to things that are unchanging and therefore by definition old. Antiques, listed buildings, preservation orders. Bits of towns will become islands of eternity to be maintained the same for ever, surrounded by whirl-winds of concrete new developments that are forever being pulled down over and over again. Ancient trees – or even those 100 years old – will acquire ever-increasing respect, together with unspoiled moorlands and woods. Old houses, more and more unsuited to an ultra-smart age of wired, intelligent low energy buildings, will continue to be popular, for those who can afford running them and the cost of converting them to fit third millennial building regulations for heat loss and carbon-use reduction.

Building a better world

Building a better world will be a dominant theme in future: corporate citizenship will attract huge attention as well as the need for workers to feel they are doing something worthwhile. Many people are realising that there is more to life than selling. There is more to life than managing. There is more to life than working. In fact there is more to Life than life itself. What will I leave when I die? How will my children remember me when I've gone? I was taking a seminar of senior executives of a major financial institution recently. There was a manager there from New York. 'I'm torn right now,' he said. 'I need to be at home with my 14-year-old son who's about to fail his math.'

He described how he argued with his son over work.

'You'll never get on if you don't learn your math.'

'Why should I?'

'You won't be successful.'

'Don't care.'

'You won't be able to get a great job like I have.'

'You can stuff the job. Look what it's done to Mum and to me. I don't want a job like that. I'd rather sweep the streets.'

The father was shocked, not only by his son's negative attitude to school, but by all the stored-up resentment his son felt over the times when Dad had left for work early, come home late, or been travelling yet again. This was a voice of a post-millennialist reacting against pre-millennial values, the belief that more means better and progress means money, money means happiness. But it may not. Hardship carries misery, but so can wealth.

Relationships are all you have left

Agony columns are full of advice about how to have happy relationships. In a fractured, increasingly disordered and fast-moving world, long-term relationships are going to matter more. One sign of success for tomorrow will be to be living happily with the same person for a long time. It shows either that you made a great choice with your partner, or that you were a highly desirable mate to be able to attract such a remarkable partner, or that you are a great partner yourself. What's so smart about a string of failed marriages, shacked-up arrangements or temporary flings?

There is a fundamental human need for security, for some things at the root of our being that do not change. Most humans cannot cope with complete and continuous changes in all areas of their lives without becoming at risk of emotional disorder and inefficiency, as more and more resources are mopped up coping with everyday life. Change is a major cause of stress: whether moving house, job, having a child or getting married or divorced. Statistics show that married employees are healthier and often more successful, and so are their children. This kind of data will go on accumulating, bringing new social conventions.

Expect to see a whole new relationship industry move out of the agony columns and marriage guidance centres and into government- and company-sponsored support and advice schemes. A workforce with happy, stable relationships saves government and employers through increased productivity and lower social support costs.

THE STRUGGLE FOR BELIEFS

In the light of all we have seen, it is no surprise to find that in the early stages of this new millennium that there is an intense, growing hunger for spirituality. In the 1960s the great debate was between those who believed in God and those who were atheists. Atheism has all but died in many secularised western countries. Now the great debate is not over whether you believe, but what you believe in. Globally there is a rise in fundamentalism of all kinds, a growth in the numbers of passionately devoted adherents to a particular religion. Part of this mass discovery of faiths (many religions are growing) has been the beginning of a new, third millennial morality.

Faith in anything, anyone. Faith that causes a peripheral member of the British royal family to crouch under a plastic pyramid because she believes it is a source of power. Faith that causes ordinary men and women to hug trees in local parks. Faith that causes intelligent people to study full-page spreads of personal advice based on the position of the stars. There is a small section of the western world that will not make an important decision if Mars is wrongly aligned with Jupiter, whether it's to buy a house, sell shares or accept a new job. Faith is everywhere.

There has been a wholesale rejection of the scientific, logical, rational model of the world that reduces all of existence to fixed, predetermined and mechanical systems. Thus doctors are having to struggle with a new generation of patients with serious illnesses, who throw modern medicines away and opt instead for alternatives which many doctors regard as unproven, untested and with little or no scientific basis. The modern medicines they reject in favour of ancient remedies have often been through years of rigorous field tests in different countries with the cumulative experience of thousands of patient years.

More than 17 million people in Britain alone now use alternative medicines or therapies, aromatherapy and homeopathy being the most popular. So some third millennial patients have more confidence, when it comes to health, in the alternative lifestyle than in the scientifically proven. Expect laws to tighten, requiring companies to verify claims made for improvements in health. This will intensify a culture clash between those who feel that scientific methodology is not a valid test of 'whole person medicine', and those who insist on 'objective' scientific data. But even the most hardened physician accepts that faith in the doctor or the treatment is vital to success.

Growth of Islam and Christianity

The great world religions are continuing to grow rapidly. While the world population grows at 1.7 per cent a year, Islam has been growing at 2.9 per cent and has 1 billion adherents, by 1997 representing 20 per cent of the world. Christianity has also been spreading at 2.7 per cent a year, with 1.7 billion adherents and 33 per cent of the world population. Explosive growth is being seen in many of the poorest nations, with a rediscovery in the richest nations and a reassertion of the validity of the Christian faith at an intellectual level.

The Church is growing fast in all former communist countries as well as in China, where millions have found faith since the 1950s despite a history of severe persecution. In Korea there is at least one church congregation of more than a million members. In Argentina over the last decade churches have sprung from nothing to number many thousands of people, and the same has been happening across most of Latin America. Africa has seen extraordinary growth in church attendance. This renewal of spiritual energy will have a very long-term impact which is already beginning, although we are in the very earliest stages.

Even in Britain, a country where public displays of religious fervour have tended to be frowned on, politicians in the last two elections were seeking to outdo each other in public Christianness. Many have recently made a point of announcing that they go to church and that their faith has led them directly to the party they now support. In the 1980s it would have earned them no favours to use such language, but now many politicians sense an electoral advantage through being seen to be sincerely Christian, in a world increasingly dominated by sleaze. This public identification with traditional Christian faith is very significant in an age when so many had written off the impact of the Church.

Politicians are populists and experts at sensing a change in public mood. They clearly sensed that the mood of the nation was changing, from a collective scepticism about faith to a fresh hunger for it and genuine respect for those who had it themselves.

The nation certainly has been shifting. Between 1994 and 2001 an estimated one million British adults including myself attended a 12-evening course (and most of them a residential weekend) introducing them to the basics of the Christian faith. These Alpha courses swept the country after being piloted in a large Anglican church in central London: Holy Trinity Brompton. Over the same period it was not uncommon to see up to 70,000 people gather in the open air in London to pray for the nation, walking through the streets with colourful banners as part of the March for Jesus. This London-led phenomenon quickly spread with annual simultaneous marches happening in dozens of cities. On one day in June 1996, around 12 million people marched in over a hundred nations across every time zone.

Then there are annual week-long residential Christian conferences, now attended by well over 250,000 people in Britain, who spend millions each year to hear the best teachers and experience a dynamic across the denominations that they can never find in their own places of worship. All this is surprising to people who drive past redundant church buildings, or wander into vast cavernous churches filled by just a tiny handful of the elderly. There is still a demographic bulge, which means that while attendance is growing strongly in the younger age groups it is not completely offsetting a loss through death of a much older generation. But something is changing.

Religious politics

Over the next decade we will see growing political movements powered by religion, for example Hindu fundamentalism in India and Buddhist fundamentalism in Sri Lanka and farther east. It has already been happening for a long time in the Middle East with

Islam. In 1980, Islam's fifteenth century heralded the Iranian revolution, Afghan religious leaders declared war on Russian infidels and religious extremists killed Anwar Sadat, president of Egypt. The Islamic political movement is now spreading far and wide.

Kashmir is one area where Islamic militants are a strong force, while 15,000 Turks recently protested against their government's plans to curb Islamic influence in education. Water cannon and clubs were used to break up riots.

In the 1970s or 80s there were posters everywhere in Britain with messages such as 'No to Conservatives, Labour or Liberals, yes to Socialist Workers Party' – the SWP was and is on the extreme left. Today those posters are more likely to say 'No to the rest, yes to Islam'.

In the meantime, Arab states continue to live in uneasy tension. While Iran appears to be softening its attitudes and fundamentalism is ebbing in Egypt and other countries locally, attempts to stamp out religious fanaticism in Egypt have cost at least 1,000 lives since 1992. While the extremes of moral code and dress are being challenged and replaced by a more moderate form of fundamentalism, it would be a grave mistake to think that Islamic politics is on the wane. Muhammed Khatami's presidency of Iran may on the surface appear to be heralding a new, moderate form of Islamic rule, backed by an overwhelming share of the 66 million voters, but instability remains.

In 1996 the US passed the Iran-Libya Sanctions Act, punishing any firm investing more than $40 million in Iranian or Libyan oil. Tension between the US and this region has continued over the issue of sanctions against Iraq, which some suggest may have resulted in deaths of over half a million children. One positive step in British perceptions would be the lifting of the late Ayatollah Khomeini's *fatwa* ordering the murder of British writer Salman Rushdie for allegedly blaspheming against Allah in his writings. The fact is that the Arab national and tribal psyche is far removed from the western mind.

Expect huge efforts to promote greater understanding between Islamic nations and those with a Judaeo-Christian heritage. Expect similar efforts to obtain a lasting settlement between Israel and the Palestinians, in a wider search for global stability.

Christian politics

Christians are also becoming politically aggressive, not only through single-issue campaigns over issues such as abortion and euthanasia, but through organising the beginnings of new parties such as the Christian Democratic Party. The Pope continues to encourage Catholic action – for example over abortion. He continues to have huge pulling power – drawing crowds ranging from 50,000 in a Baltimore baseball stadium to over a million in huge open-air gatherings in other nations.

Rev. Pat Robertson, champion of the religious right, is typical of many to come. So is Patrick Buchanan. When Buchanan ran for president many were confused. Pro-gun and anti-abortion, pro-workers' rights and with a Christian moralist label – both left- and right-wing. He didn't make sense in any old-style framework. But Buchanan was a sign of the future: a man driven by a cluster of controversial issues which made perfect sense one by one to him and many others. President Bush has made no secret of the strength of his own Christian faith.

Jack Kemp, Republican vice-presidential nominee, addressed the Christian Coalition's 'Road to Victory' conference in September 1996: 'I could not be here today as a candidate for vice-president were it not for the prayers of the men and women in this room and across the country. How can you be a Christian and not be involved . . . ? We're not trying to build the City of God on earth . . . But we must on earth work for peace and hope and order and equality in the City of Man.'

Expect to see further contrasts, particularly in the battle-ground over schools in the US – where school prayers are banned, but where Christian schools will go on growing. Expect to see America divided by new faith-government partnerships, using the Church as a delivery mechanism for drug rehabilitation and a host of other social programmes funded by the state.

Whether you are a follower of Jesus as I am, or of Mohammed, or of Buddha or the patterns in the stars, whether you are a devout Jew, believe in Karma or reincarnation, or some other life-force, or in nothing at all, spirituality will be a central issue shaping values, politics and people movements, a dominant influence for the next fifty to a hundred years.

CENSORSHIP

Censorship and the media

Islam is highly insistent that the Koran should not be defamed. Christianity has been more docile in the past but is becoming less so. Hence the largest ever protest against the release of any film in August 1988 with the release of *The Last Temptation Of Christ* at Universal City, California.

The protests were dismissed in a statement: 'No one sect or coalition has the power to set boundaries around each person's freedom to explore religious and philosophical questions.' Yet the same company had earlier decided not to film Salman Rushdie's *Satanic Verses*. This kind of double standard will not be sustainable in the future.

The media always tend to push out the boundaries and hence are first in line for flak. So animal rights activists attacked Disney in 1990, demanding that 'anti-wolf' suggestions be removed from *White Anger*. Disney was also forced to print a disclaimer that 'there is no documented case in North America of a healthy wolf or pack of wolves attacking a human.' Even the cartoon *Beauty and the Beast* attracted similar pro-wolf protests. More recently, Southern Baptists initiated a boycott of Disney over benefits offered to gay partners and for allowing their parks to be used for 'Gay Days'.

Expect plenty more big fights between religious groups (especially Christian ones in the US) and the media. For a variety of reasons, there is a huge gulf between those in the entertainment industry and the general public when it comes to religion.

While 78 per cent of the US public pray at least once a week and more than 40 per cent attend weekly services, a survey of 104 of the most influential leaders of creative TV in the US found that 93 per cent seldom or never attended a church service and 45 per cent claimed no religious affiliation whatsoever.

Even more striking, many of those who shape popular culture have personally rejected the church. The same survey found that while 93 per cent said they had received a 'religious upbringing', only 7 per cent were currently 'regular participants' in church or synagogue services. Thus the culture and conscience clash will continue, not as a deliberate conspiracy against belief but as an inevitable expression of this huge gap in world-view.

Censorship and the Net

With pornography and gambling together accounting for between 10 and 30 per cent of all Internet trade, Net regulation will be a key target. Yet the Net also helps guarantee freedom. Look at the recent victories by opposition parties in Serbia before the collapse of the Milosevic regime. An authoritarian government controlled newspapers, radio and TV, yet students managed to spread their message on the Net, building international support, and eventually gaining power.

Censorship and link to undesirable behaviour

A huge battle will be waged over media freedom in the next millennium. The liberal trend towards total relaxation of controls will be more than offset by a conservative trend towards nannying the public. Both extremes will leave the majority of the public bemused: they will want changes to reduce access by children to adult-rated material, yet greater freedoms for consenting adults. The digital age will enable both requests to be fulfilled – in theory. In practice, every control will be subverted by adults and older teenagers who are lazy, or irresponsible. Controls will also be undermined by lightning advances in the Internet which will ensure that for several years to come any child who is sufficiently knowledgeable can access just about any film or other media he or she wishes. Expect the censorship issue to grow in importance with ever greater worries about the effects of over-exposure on a rising generation. Then will come a backlash, spearheaded by the Islamic community and Christian alliances, seeking global agreement as far as possible.

Such a religious alliance will be opposed by predominantly secular groups with claims that media output is pure fantasy for the most part, and incapable of altering behaviour. More importantly, the claim will continue to be made that actions such as rape or violence are entirely unrelated to media output. The counter-claim will grow stronger by the month, that the media has a mild influence on almost everyone, a significant influence on a few and a major effect on a small but all-important minority, who go on to commit major crimes.

Television has only been with us to any degree since the 1950s and relatively uncensored TV for less than a generation. Video recorders have been the property of the masses for even less time. Society has yet to come to terms with these media channels and the verdict will take until 2010 or longer to come, but when it does it may spring with a vengeance, with powerful government backing in several continents. There could be huge economic pressures on broadcasters, forced by advertisers who are themselves under pressure from consumers and ethical investors.

In the meantime expect to hear more voices like that of Janet Daly writing in the *Daily Telegraph*. She wrote that even if the link were unproven, 'it is wrong to base a species of entertainment on terror and the inflicting of pain. That is all that needs to be said and that should be enough.'

Right in the middle of a puritan swing expect to see strong liberal voices arguing for yet more relaxation, talking about a return to the Dark Ages.

Our world will become increasingly one of extremes

Elections will be won and lost on issues like these. We all want a market economy, but at the same time think that there should be justice, fairness and equality of opportunity in a compassionate world where health and education are available to all. But there's more to life than these things alone. We could have all these – as most people to a large degree already do in many nations – yet find we are in some kind of living hell with dark forces (some would say market forces) ripping the heart out of community life, and millions of miserable people.

Media influence is undeniable. The whole advertising industry is based on the fact that media messages change what people do. Expect a growing number of cases where there are widely publicised links between a gruesome slaying and obsession with a certain film or TV character. Expect the chorus to grow louder for controls, and also a rise in frustration levels as people discover that in a

globalised world, international agreement is needed to regulate satellite channels in particular, as well as Internet TV.

Media ownership

Expect increasing protests over the centralised control of globalised
media empires able to pump out vast amounts of propaganda on
particular issues when needed. People like Rupert Murdoch will
come in for a lot of criticism as their highly successful empires
control more and more traditional media channels.

However, the voices will have to contend with the argument that
the increasing glut of choice of channels, together with the Internet,
also dilutes power. Indeed, politicians and advertisers alike will have
to work harder than ever in tomorrow's world to catch a nation's
attention in a way which would have been comparatively easy in
the 1960s.

Battles over software will be at the centre of the future media wars,
with decoder boxes at the heart. Which decoder do you have? Whose
software are you using? Which cable network serves your city?

A NEW WORLD RELIGION?

Expect to see all major world religions continue to reinvent themselves, as their traditional teachings are reinterpreted in a very different age and culture from any ever seen before. The roots of the
Christian faith have remained virtually unchanged over 2,000 years
but expressions, understanding and practice have varied greatly. For
example, in Christianity expect to see new mixtures of the radical
(attempts at ultra-cultural relevance, youth churches and other
experiments) with the traditional and mystical. Tradition fuses our
present experience with the living spiritual echoes of worshippers
over centuries. Symbolism will become very fashionable as a means
to express depths beyond the trivia of words. Life-changing faith
will spread, adapting to culture, then altering it significantly.

A totally globalised world will create a 'market' or a vacuum for
a new world religion, which will feed into the aspirations of the
third millennium. We could well see a world-recognised prophet

emerge over the next few decades with charisma, dynamism and teachings which capture the global imagination. The biggest issue will be truth: is there such a thing? While the great religions such as Judaism, Christianity and Islam are rooted in historical events and proclaim timeless truth about an unchanging God, offering exclusive understanding of him, new age beliefs in the late twentieth century have borrowed heavily from some aspects of Hinduism, which emphasises a more general approach to truth, and a fluid ethical framework without absolutes.

A new world religion is unlikely to be just more of the same, in other words it will not be just a mish-mash of beliefs centred around a conviction that there is no such thing as absolute truth.

In a constantly changing world certainty about ultimate issues such as personal destiny becomes increasingly important. That is the appeal of fundamentalism. Expect therefore that a new world religion will be marked by dogmatic teaching and a claim of exclusivity and superiority to all previously understood truths about God. Expect a prophet of such religion to offer 'the final revelation', the missing pieces of understanding of perfect knowledge that humankind has not before been ready to receive. Such a prophet will promise that humankind is 'coming of age' and is only now able to receive the truth. The claim will be that all the great religions pointed in part but did not provide the complete picture. Such a prophet could sweep millions of adherents of other religions, including Christianity, into a new religious movement.

A NEW WORLD ORDER

With so many problems in the world, will we see a new world order? Is there a way of rolling together the authority of 222 governments and thousands of sub-national governments worldwide? The British Empire was built on the belief that world government could bring huge benefits to humankind, before it dissolved into a Commonwealth of 51 states. In one sense a new world order is growing by default, as we have seen, for example, over the proliferation of environmental treaties, and the global fight against terrorism. Yet institutional rule is weak and weakening.

More international treaties will create global control

The more international treaties there are, the more an informal global government emerges. Expect more agreements like Kyoto (global warming), START (Strategic Arms Reduction Treaties), the Chemical Weapons Convention and the Open Skies Treaty. Similar agreements on a wide range of issues will attract more than 170 nations as signatories. The United Nations was founded to 'save succeeding generations from the scourge of war' and 'to reaffirm faith in fundamental human rights, in the dignity and value of the human person.'

The real trouble is that, as we saw over the Iraqi conflict, UN member nations are unclear what they want the UN to do and how it should perform those roles. Peacekeeping has become politically and practically dangerous. The emphasis has switched to building economic security. The UN council is unlikely to embrace another Iraq or Somalia or Bosnia with enthusiasm, but countries which stop in-fighting will be offered plenty of peace-building incentives. The UN will help on elections, the judiciary, education, health, government infrastructure, agriculture and trade.

One of the roles the UN performs best is the preparation of the world's weakest economies for foreign and internal investment. The great problem is that the 185 nations in the General Assembly have widely differing opinions and agendas. Each is clear about what it wants but there is no agreement.

Expect the fall-out from the Iraq war to be profound, with huge efforts made to try to re-integrate the US into a consensual approach to world affairs. Expect the US to pursue dialog while retaining right to independent action on all matters of national interest: a stance which will provoke growing anger and resentment around the world. But the trend is clear: global governance is going to be essential to our peaceful and prosperous future. The unprecedented spirit of collaboration that has emerged since the collapse of communism will deepen despite crises and setbacks. A key global challenge will be finding a solution to the Israeli/Palestinian issues of security, justice and peace.

Code of conduct

Despite differing positions on many issues, there is growing agreement on an international code of conduct. Indeed, international trade is impossible without it. World trade requires basic principles: integrity, honesty, obligations, mutual respect.

International courts

You cannot trade unless there is trust that both parties will honour the agreement. There are rules to the game, and those who play are expected to keep to them. The Nick Leeson episode and the collapse of Barings Bank showed that there were two versions of the rule book: official and unofficial. But rules need policing.

Voluntary codes of practice and self-regulation will not be enough. International laws will be needed. While individual nations may reform their own legal systems, another level will be needed and created. The beginnings are already here: for example the prosecution of national leaders in one country for war crimes, by a court comprised of other countries' representatives.

Expect to see a growing number of countries sign up to an international court, able to hear cases which are almost impossible for any single country to deal with. At present attempts to create global law for a global village are stopped before they have even started in many cases by extradition procedures, which are usually activated only if both countries accept that a crime has been committed, and respect each other's legal systems.

Regional law courts are well established, in the EU for example. Expect supra-regional courts to be dealing with a wide variety of international crime cases by 2010. History shows that law and order is imposed most rapidly where there is a lack of it. Therefore these new powers will be agreed by nations as a matter of urgent necessity, faced with their own impotence.

Cyberspace is a country needing government of its own

One interesting area will be laws governing the Internet and cyberspace. It is already becoming obvious through disputes over the naming of sites that nations are losing control to a new nation, to a new territory altogether. As we have seen, whole communities are developing a complete trading environment and cyberspace is almost tax free.

Until early 1998, no one in the world could get a name approved (i.e. an address on the web) without a US agency authorising it. Endless rows broke out as companies in different parts of the world with the same name began to fight to use their name in the cyber-village. Which company should win? The first to register? But is that unfair when a tiny company has won global cyber-rights to the name that is also that of a multinational known the world over? Should the US always decide? Clearly it was a nonsense to give so much power to one nation, but what would replace it? From where should cyberspace be run?

At present, cyberspace is run by a benevolent dictatorship, created mainly by US agencies. This cannot and will not last, nor will any other benevolent dictatorship of non-elected, unrepresent-ative authority. Expect calls for democracy in cyberspace with elec-tronic votes for every cyber-citizen (e-mail user) who registers to vote. Then we will see elected cyber-government, with legislative powers backed by a moral imperative. Of course those powers will only operate inside the cyber-world.

GLOBAL GOVERNMENT

In summary then, expect to see various expressions of global govern-ment emerge, and co-align into the first stages of a new world order over the next few decades, with components derived from all the above. The beginnings will be hardly noticeable, but the technology of tomorrow could give it remarkable strength. Globalised struc-tures to regulate a globalised world – although at every stage con-struction will be slowed by tribalism and other radical forces.

Expect periods of intense negotiation to define global ethics in more detail, whether attempts to create a total global ban on human cloning, a response to international terrorism or ethnic genocide, limits on global monopolies, or world agreements on slavery, child labour, work practices and other issues of human rights and res-ponsibilities. Many issues will be polarised between emerging and developed nations. All these debates will be increasingly influenced by a post-millennial rethink in wealthier nations, in a generation no longer impressed by speed, urbanisation, material wealth and globalisation. A generation which itself is becoming radical, ethical and spiritually aware.

CHALLENGES TO MANAGEMENT

Where do your ethics come from?

◆ What are the key ethical issues facing your company in the next decade?

◆ Where do your corporate ethics come from? Shareholder views? The board? The chairman? Each region or department? Informal chats with colleagues at conferences?

◆ Do you have an ethical forum, committee or think tank?

◆ How do you decide when company ethics need to change with the times, for example over corporate governance.

How well protected are you against civil litigation?

◆ Class-action suits are expensive to deal with and can tie up a major corporation for five years or more, with a damaging effect on company valuation and stock prices.

◆ How well protected are you against group actions?

◆ How sensitised is your legal department to which angry groups might take your company on in the third millennium?

Political correctness

◆ Who decides corporate policy regarding political correctness, e.g. the use of language, sexist terms such as 'chairman', or are these things decided chaotically and inconsistently at hundreds of different levels?

◆ Are you culturally sensitised to where it really matters and where it does not?

◆ Have you resolved the globalisation conflicts where a global approach may be disastrous in, say, Saudi Arabia on the one hand and the US on the other?

Motivation

◆ How able are you to motivate staff who are not so interested in more money?

◆ How are you going to retain the best who don't need you and want to work under their own terms, perhaps contracting to you part-time or more flexibly, maybe for less cash?

◆ Is your personnel policy geared up for this growing social revolution?

◆ Have you fallen into the trap of encouraging a culture where people who are at work from 7 am till 10 pm are seen as 'better' than those working fewer hours?

◆ How can you improve morale and job satisfaction, helping staff to feel that in some small way they are helping to build a better world?

◆ How can you help create a feeling of family, of belonging?

◆ What about the balance between work, leisure and family in your own life – are you a model worth following?

Reaction against constant change

◆ Are you ready for a world where constant change is seen as far less attractive than stability and constancy?

◆ What will that do to your corporate culture?

◆ How can you help people find security in things that do not change, to help them cope with changes that do have to be made?

Respect for relationships

◆ How friendly is your company to men and women who have just got married, or have young children, or who are fighting to save their marriages, or who have sick and dependent relatives at home?

◆ Is your corporate culture likely to increase marital happiness and child wellbeing – or contribute to break-up and distress?

◆ Does your company policy reflect the importance of happy home life in obtaining the greatest productivity from employees?

Spirituality

◆ Is your company sensitised to changing cultural influences in many nations as a result of a revival of commitment to religion?

PERSONAL CHALLENGES

Ethical challenges touch sensitive areas in our personal lives. Here are some important questions that many senior executives are asking, as they grapple with life in an increasingly fast, urbanised, tribal, globalised, radical and chaotic world.

Giving your values high priority

◆ How are you going to thrive in a world which is fast, urban, tribal, universal and radical?

◆ How do you want your future to be different from the past?

◆ How much time have you given to working out what motivates you, what your own values are?

◆ What holds your life together? Is that reflected in the workplace or is there a danger of leaving your values behind?

◆ How motivated are you by your current work situation?

◆ Do you work for an organisation which reflects your own values?

◆ If not, how important is that to you?

◆ Do you need to look for another job?

◆ When are your personal values more important than promotion, e.g. time with family, less mobility, honesty and integrity?

Upwards or downwards mobility?

◆ Is your priority for the next five years upwards or downwards mobility, or to stay about the same?

◆ Have you considered creative opportunities to work in different ways – for example with job sharing?

Consistency

◆ Is your own management style consistent with your values?

◆ Is your lifestyle consistent with your own values?

Stability

◆ In a world which is constantly changing, which parts of your own life are going to stay the same?

Relationships

◆ Is there more to life than work for you?
◆ Is that reflected in how you live?
◆ How balanced are the various areas of your time and energy?
◆ How will your children remember you when you have gone?
◆ How are you preparing them for the future?
◆ Are you happy in your relationships?
◆ What can you do to improve the situation?

Purpose

◆ What is the purpose and meaning of your life?
◆ How are you helping to build a better world?
◆ Do you have a sense of direction other than merely pushing doors, seizing opportunities and enjoying what life offers?
◆ What do you want to achieve?
◆ What are your personal aims and are they realistic, achievable, measurable?
◆ How will you recognise your own success?

Spirituality

◆ What does faith mean to you?
◆ How important is spirituality to you?
◆ Is this an area you want to invest in?
◆ When work and money have gone, what will you have left?
◆ Are you giving yourself enough personal space?
◆ Do you have a sense of personal destiny?
◆ Are you spending time with people who have a spirituality you respect?

Afterword

Turning the cube – Optimist or Pessimist?

So then, these are the six faces of the future. As I lecture about the future around the world people often ask me if I am an optimist or a pessimist. Surely, they say, the future contains so many disturbing challenges that it must worry you? The truth is that I am excited about the future and pleased to be alive at this time in history. The potential of technology is intoxicating, and the moral challenges we face are considerable. The choices are for good or for evil. To fulfil our own God-given destiny or to abandon the great design for living, the road to chaos and destruction.

It's impossible to keep all six faces of the future in view at once, and some are more important than others. But all are vital to keep in view from time to time. Some are related as pairs and seen together: Fast and Urban, Radical and Ethical. Tribal and Universal are opposites and hard to view at the same time.

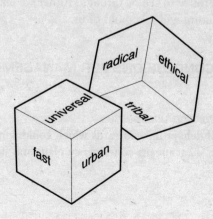

The interesting thing about the cube is that it creates two different worlds. The world most executives live in most of the time is fast, urban and universal. A rapidly changing, globalised village. However, there is another side, a world which is tribal, radical and very ethical. It may involve fewer people, but how many tribal, radical and ethical people do you need to change a society?

When I pose that question to high flyers from all over the world they almost always come back with a similar answer: very few. You don't need many radical activists in a population to change its values, to alter government policy, to make a difference to how corporations behave. Most CEOs, chairmen, board members and senior managers would say that probably less than 2 per cent of the population with tribal, radical and ethical outlooks could be enough to affect a society profoundly.

We can argue about the exact percentage, but one thing is clear: while it is right for organisations to focus mainly on the obvious and immediate challenges of a fast, urban and universal world, they also need to keep turning the cube (not tossing it at random) but from one side to the other.

As someone who shares a common desire to build a better kind of world, to challenge values, to provoke justice, fairness and a compassionate use of technology, for the benefit of the whole of humanity, I find it very encouraging that the doors of influence are so open. I enjoy the fast, urban and universal world but am disturbed by its challenges. Never before in human history have we so needed ethics as much today to see us through. Either we take hold of the future or the future will take hold of us.

TEN CONCLUSIONS FOR MANAGEMENT

1   Prepare for the unexpected
The future will deliver us wild cards which could win us the game or wipe us out. That means contingency plans and flexibility.

2   Faster reaction times

Everything needs to be geared to rapid response from top to bottom. If you think things are changing fast, get ready for double and treble the present speed of change.

3   Flatter structures

Pyramids can't cope with post-millennial life. Either get flatter or get smaller. Build associations, franchises, partnerships, co-operative alliances. Decentralise and empower.

4   Teams and partners

Third millennial life is too specialised to have everything done in-house. Teams and partners keep everything alive.

5   The global village

Globalisation has only just begun. The millennial generation is growing up in a global village that we have only seen the foundations of. A key challenge will be to shrink management lines geographically using technology. Those who can't cope with virtual communications just won't survive.

6   Cultural sensitivity

Globalisation will mean more cultural sensitivity not less. The oldest mistake in the book is to think that people think like you because they speak the same language.

7   Investing in technology

Information tech, biotech – use it, control it, make it happen. Select people who like technology, who are excited by it, who have the creativity to adapt around these new tools.

8   Creating family

In this fractured and dislocated world people still spend more time working for you than doing any other single activity, so take care of them. Make them feel special. Make them family, with a sense of identity, value and belonging.

9   Purpose and meaning

Help your team find ways to feel that they are building a better world. Who wants to spend their entire lives doing something that adds no value to anyone, has no purpose and no meaning?

10   Leadership will be everything

No amount of committees, strategies or parallel plans will help one jot without dynamic visionary leadership. So how do you know if you are a leader? How do you spot leaders under you and over you? Ask one simple question: Are people following you? Who are they following?

Leaders don't lead by position, they lead by inspiring trust and confidence. Leading through fear of punishment invariably results in disaster, with resentment and rebellion sowing the seeds of destruction. Leading through dynamic vision and motivation invariably results in energy and progress.

TEN CONCLUSIONS FOR INDIVIDUALS

1   Prepare for the unexpected

Your whole world is going to be changed – now is the time to prepare. You have already begun that process by reading this book and working through some of the action points. You can help shape and build a better world.

2   Plan to react faster

Some of the greatest opportunities can flash by in a moment. That means thinking things through now, and discussing possibilities with others – for example your partner. Remember: take hold of the future, or the future will take hold of you.

3   Invest in technology

One of the best investments you can make is in a powerful personal computer with high-speed Internet access. Get familiar with what networks can do for you and keep watching – the digital society is growing faster than you think.

4   Keep well informed

Those who stay ahead of the future will be exceptionally well informed. So how do you manage this without adding to information overload? Read a weekly summary such as the *Economist* together with a couple of quality daily newspapers. Make sure that you regularly skim-read a couple of popular computer magazines.

5   Stretch your horizon

Take opportunities for executive training – just the experience of meeting others will stimulate fresh thinking, as will the presentations.

6   Think laterally

Most people are blind to their own potential. Employers in future are going to need some very unusual combinations of skills and backgrounds. So keep an open mind about the sort of jobs you could go for. Keep building on what you have but keep broadening too. The next step up could involve a sideways move.

7   Make time for people

In ten or twenty years' time the world will have rushed you by and all you will be left with are memories, money and relationships. But without relationships, you have no one to share memories with or to enjoy what you have. At the end of life, relationships are all you have left. Invest in people.

8   Be who you are

With so many conflicting pressures and events, be who you are. Don't let the system clone you into conformity. Stick by what you believe and what you know to be true. People will respect you for it. Take time to reflect. Explore your own spirituality and faith.

9   Enjoy today

You are the most important person affecting your future. Life can only be lived once. Be kind to yourself. Enjoy each day. Seize the moment before it fades. Today is the day of opportunity.

10   Celebrate the past

Celebrate the past, with its highs and lows, the good and bad times, the triumphs and disasters. It has all shaped the present, made you what you are, and your past will help you understand your future.

Take hold of your future – or the future will take hold of you.

Appendix

Measuring your FUTURE

So how can you measure up your organisation against the Six Faces of the Future? How can you assess your own strengths and weaknesses in tomorrow's world? I am grateful to David Stanley's imaginative help in developing this simple 'Star Treatment' method of Futuring your organisation. The six-faces of the cube are, as we have seen earlier in this book, weighted in different ways according to time, place and industry. But if we squash the cube flat and make a circle we can begin to make a Future-Chart as a measuring device.

The technique is simple and fast: mark on each of the six lines how strong you think your organisation is in the six key areas FAST, URBAN, TRIBAL, UNIVERSAL, RADICAL and ETHICAL. When the points are joined up you get a unique shape. Every industry has its own preferable shape – for example an internet bank is strongest shaped as a sword, an ethical investment fund is strongest as a butterfly.

When you and your team have FUTURED your own organisation – do the same for your competitors. The final stage is to re-FUTURE your organisation as you would like it to be in five years' time. This then can help locate areas where the most change is needed and enable a focused analysis of key areas for transition. Of course, this is an empirical 'feel' and a more comprehensive FUTURE audit is a major task, based on a comprehensive survey, using a calibrated series of formal questions and answers to identify where the organisation is at each element on the chart.

See typical industry profiles below – used as worked examples. Of course every executive in every company will have their own view as to the FUTURE rating of their organisation, but the discussions generated are very valuable in identifying strengths and weaknesses.

THE BASIC GRID

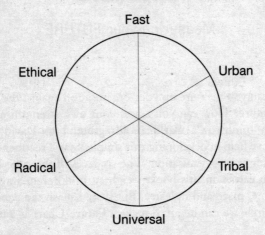

The typical and successful e-bank is strong on Fast and on Universal, but weak almost everywhere else because the other faces are less important, and energy directed there may adversely affect the core business.

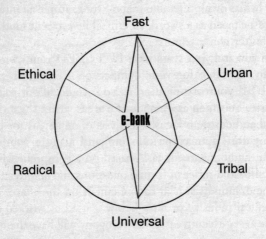

In contrast, an ethical investment fund is usually based in one country – partly because ethics vary so much with culture. So it scores high for Tribal, and also for Urban as it benefits from factors such as the ageing population. Of course it also rates high on Radical and Ethical, but not on Fast because ethical funds are usually by their nature cautious, reflective, safe and secure in approach.

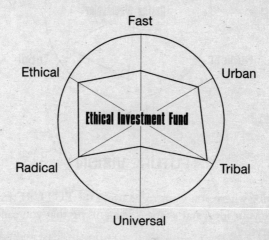

The coffee wholesaler is in a mixed position typically with a range of complex issues to face. Coffee drinking scores high under Urban as a social institution, dependent on demographics and fashion. It is also a Tribal drink, since it is usually shared in a family or group of friends or business colleagues. Coffee is a global business – the second largest commodity traded in the world after oil, so scores high on Universal. However coffee scores low (usually) on Radical, since activists are successfully damaging the image of the coffee industry as exploiting the poor. It also scores low on Ethical, since the industry seems unable to form a clear ethical framework for what it does, adding to the Radical vulnerability.

Although the initial process is informal and can be carried out rapidly in the context of a workshop, a rigorous methodology can be applied to the process. The first step is to identify the vision-holders and influencers in the organisation and to bring them together.

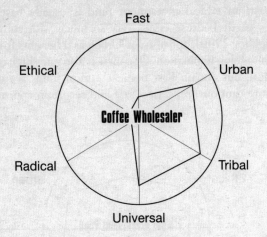

FUTURE VISION

- Describe where you want to be using the FUTURE framework.
- Create your ideal star shape – recognising that you can't be best at everything.
- Compare with star shapes of competitors.
- Workshop with key decision-makers to identify 5 key factors for the company's future success, coming out of each of the six faces. Out of these 30 issues, select 10 as the most important and split into two: high and medium impact.

CURRENT STATE

- Electronic surveys of opinion-shapers and other target groups are analysed to see how well the company shapes up to the challenges as identified. e.g. How good are we at encouraging flexible home-working?
- These are then combined with other data gathered from various sources to help establish key target areas for action, which can then feed into an effective change–management process.

- Sharing the results of the FUTURE process and the emerging new vision, plus the need for action.
- Follow through with measurables and strategic planning/action.
- A result should be that the FUTURE shape of the organisation changes.

Index

enquiries@globalchange.com   00 44 7768 511 390

Trends analysis and organisational development

Global Change Ltd is a management consultancy company, providing global trends analysis and forecasting, as well as assisting corporate change and executive development. It was founded in 1996 by Chairman Dr Patrick Dixon.

Services for Fortune 500 companies/multinationals

- Global forecasting – vital trends which affect corporate survival: globalisation, economic instability, market changes, production and distribution, technology, computers, networking, virtual offices, socio-demographic changes, biotechnology, science, medicine, financial services, tribalism, political changes, single issues, lifestyle changes and global ethics.
- Management of change – challenges to management created by ever faster events: dynamic workshops/think tanks/conferences.
- Senior executive training – taking a broader view, decisive management, coping with accelerating demands.
- Long range vision-building at Board level – five to ten years and beyond.
- Corporate research – ultra-fast report generation tailored to specific areas by expert teams, backed by latest technology including intelligent agents (information robots) with comprehensive analysis and clear executive summaries.
- Information technology consultancy – Internet strategy and implementation – maximising cyberspace profitability in a market

which is changing at seven times the speed of traditional business.
Web-site development.

State-of-the-art multi-media presentations
- World class presenters using cutting-edge technology
- The medium is the message when talking about the future
- Experience tomorrow's world today
- Digital control – Full-motion video sequences – Virtual reality simulations
- Video-conferencing – Computer animations – Multi-screen
- In-house design and production including video-compression and digital editing

Media response unit
- High impact, authoritative comment on global and ethical issues
- Rapid response – Background briefings – Analysis
- Television/radio features/studio
- Documentary research and consultancy

Global Change Ltd

Booking Dr Patrick Dixon, Chairman
patrickdixon@globalchange.com

Dr Patrick Dixon is available for global bookings and multimedia videoconference presentations through Leigh Bureau, the longest established speakers bureau in the United States (telephone US 1 908 253 8600). Initial discussions regarding audience, content or style should be directed to them or you can e-mail Dr Dixon. Each presentation is unique, customised to the precise requirements of each client, using the Global Change research team and Dr Dixon's global experience.

Global Change Ltd

List of recent clients
ABN AMRO
Abu Dhabi Police Directorate (UAE)
Accenture – partners
AIC Carriers World – keynote speech
American Apparel Manufacturers' Association
American Management Association
American Society for Training and Development
Arthur Andersen – partners
Aviva
Bank of Ireland
BASF
BBC
Britannia Building Society – senior team
BT – BT Openworld
BUPA
Carlton Television
Clariden Bank – board
Compaq – European executive board
Concourse Group
Corenet Global
Credit Suisse Private Banking – major client
Credit Suisse Investment Banking – executive board/senior team
CSC Consulting
Danish Post Office

Dubai Ports Authority
EFMD (European Federation of Management Development)
Etisalat Telecom, United Arab Emirates – consultancy
European Coffee Federation
European Institute for Research Management (EIRMA)
European School of Management and Technology
Exxon Mobil
Food Business Forum – global (CIES)
Fortune – CEO Summit
Freshfields – global partners keynote
Georgia Technology Forum
Gillette – board and senior team
HEC
Hewlett-Packard
Hotel Ecoliere de Lausanne
Houston Energy Forum
HSBC
IBM – client event
IMG
Institute of Management
International Mutual Funds Institute
Jebel Ali Free Zone (Dubai)
Kraft Jacobs Suchard
Knight Frank – partners conference
Linkage
Management Center Europe (MCE)
McCann Erikson – board plus senior team
Microsoft – client event
Minsheng Bank, China (World Bank Technical Assistance
    Programme)
Morgan Stanley Dean Witter
Pinnacle Communications
Pom+
Post Office (UK)
PricewaterhouseCoopers
Prudential
Qualcomm – senior executive team plus top 750 people
Regulatory Affairs Professional Society (RAPS)

Richmond Events
Roche – Switzerland and UK client events
Royal Bank of Scotland
SAir Group (SwissAir) – board plus senior team
Saks Inc (incl Saks 5th Avenue) – consultancy
Sara Lee – board and senior team
ServiceMaster – executive board
Schweizerische Gesellschaft fur Organisation
SEB group – board plus senior team
Siemens
Skandia
Smith and Nephew
Strategos
Sulzer
Sumitomo Bank, Japan
Sustainability Forum – Zurich
Swiss Stock Exchange – executive board
Tear Fund
Tetrapak – board and senior team
UBS (Private Banking, Retail Banking and Central Finance
    Divisions and Cyberbanking Stra Development for pre-merger
    UBS Group Executive Board)
Unaxis – board and senior team
United Nations – UNAIDS and UNIDO
University of Salford Management School (Keil Centre)
Unisys
Urban land institute
VHA
Vontobel Bank (Switzerland)
Winterthur Insurance
Wheaton College – board and senior team
World Bank – technical assistance team to China
ZFU
Zellweger Luwa AG – board and senior team
International conferences: World Economic Forum (Davos),
    Southern African Economic Su Emirates International Forum,
    Internet Expo 1998 (Helsinki), European Federation of
    Management Development